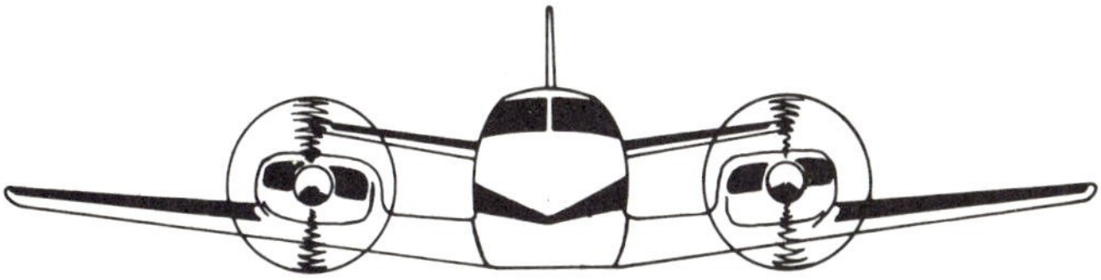

FLYING ADVANCED MULTI-ENGINE AIRCRAFT

BY PETER LERT

MODERN AVIATION SERIES

TAB BOOKS

BLUE RIDGE SUMMIT, PA. 17214

FIRST EDITION

FIRST PRINTING

Printed in the United States of America

Library of Congress Cataloging in Publication Data

Lert, Peter.
Flying advanced multi-engine aircraft.

bibliography: p.
Includes index.
1. Twin-engine airplanes—Piloting. I. Title.
TL711.T85L47 629.132'5243 80-19918
ISBN 0-8306-9697-0
ISBN 0-8306-2299-3 (pbk.)

Cover photo courtesy of Beech Aircraft Corp.

Preface

Are you a freshly minted multi-engine pilot? Or perhaps you've had the multi ticket for a while, but are planning a move toward larger aircraft? If so, then this book is for you, just as it's for the pilot planning a career as a corporate flyer, a would-be commuter captain, or even a single-engine pilot planning on getting that coveted "manymotor" rating, and wondering what lies beyond the light twin used for training. You may never plan to fly a twin larger than a Seminole, Cougar, or Duchess, yet there's no law that says you can't fly it with just as much competence, knowledge, and professionalism as the captain of some corporate gold-plater. You'll find that taking an advanced approach to your multi-engine flying will not just increase your own satisfaction and feeling of accomplishment, it'll make the flights more enjoyable for you and your passengers, and you may find that it becomes a stepping stone toward such pinnacles as the Airline Transport Pilot rating.

Reading this handbook, you'll find that I've made a couple of basic assumptions: while any activity that draws as many rugged individualists as pilots can hardly have an "average reader," I'm writing, at least to some extent, for a reader who has a multi-engine license earned in any of the light trainers such as the three previously mentioned, as well as older types such as the Piper Apache or Twin Comanche, or the Beech Travel Air. I'm sure that many of you also have some experience in slightly larger airplanes like the Piper Aztec, Beech Baron, or Cessna 310. This primer, then, will introduce you to the world of "medium" twins, 6000 lbs. gross weight is the usually accepted dividing line, as well as the "heavies," most of which, at least nowadays, are turboprops. Don't put the book down, though, if you have yet to earn your multi wings. The first chapter is a review of multi-engine basics, and even if you're strictly a single-engine type for the moment, it'll give you enough understanding of multi-engine procedures to understand everything that follows. It might not hurt you old-timers to have a look at the basics, particularly you fortunate owner-pilots who may be contemplating the purchase of one of the lighter pressurized twins. How long has it been since you've done any single-engine work, for example?

Another assumption I've made, particularly in the chapters on advanced avionics, is that most of my readers are instrument-rated. If you're not, frankly, you should be. The airplanes we're going to discuss are not only sophisticated enough, but also expensive enough that they'll often be used as business machines, rather than personal transportation. As such, you, the pilot, may be called on to get the job done now, if not regardless of the weather, at least in less-than-ideal conditions. Having a sophisticated aircraft and an instrument rating do not, of course, mean that you can take on any weather at any time (not even the airlines can do that, and besides, it's not nice to fool Mother Nature). The fact remains, though, that many of the airplanes we're going to discuss are usually flown by professionals. They're designed to be flown by pros, and even if you own the ship yourself there's no reason to be anything but a complete professional once you step forward of the cockpit-cabin divider. Moreover, most of the pressurized aircraft, and virtually all of the turboprops, don't really come into their own terms of fuel efficiency until they reach positive-control altitudes, so that any long-range flight will be, at least legally, IFR all the way.

Finally, you can consider this book a primer for the whole world of advanced general-aviation aircraft and beyond. We've already reached the stage at which the sophistication and capabilities of advanced medium and heavy twins rival, and in some cases surpass, those of the airlines. With the dawning of the (necessarily!) energy-conscious Eighties, it may well be that the fuel-efficient turboprops and big recips will play an increasingly important role in the upper levels of general aviation. While the big jets will always be a part of the scene, much of the projected growth of corporate aviation is in the medium and heavy twin area, after all, on stages of a few hundred nm's, the faster turboprops are only minutes behind the jets, while burning half the fuel! It's these planes that may well offer a great proliferation of jobs in the future. Also, when piloting seats in the jets and on the airliners become available, those in charge of hiring will be looking for experience and training in our chosen field—advanced multi-engine flying.

Acknowledgements

The author wishes to thank the following companies for materials and support in writing this book: Beech Aircraft Co.; Bendix Avionics; Cessna Aircraft Corp.; Collins Radio; Garrett Corporation; Mitsubishi Aircraft; Rockwell International (Gen. Aviation Div.); Swearingen Aircraft Div. of Fairchild Hiller; and United Aircraft of Canada Ltd.

Dedicated to all the students from whom I've learned and especially Diane.

Peter Lert

Contents

1 **Review of Basic Multi-Engine Techniques** 7
Basic Configurations—Controllability and Performance—Engine Failure at Takeoff—Dealing with Engine Failure—Fly the Airplane—Clean Up the Airplane—"Dead Foot, Dead Engine"—Feather the Engine—Critical Engine—Basic Reactions to Engine Failure

2 **Handling Large, Complex Airplanes** 17
Light Twins Vs. the Large Aircraft—Bigger Planes Give Smoother Rides—Stay Ahead of the Game—Learn to Commit Yourself—Forced to Go Aground—Handling Techniques—Weight and Balance—Loading Flexibility—Fuel Loads—Relax-It's Not That Complicated

3 **Large Aircraft and Their Systems: Complexity Vs. Redundancy** 28
Fuel Systems—Fuel Transfer Pumps—Possible Fuel System Hangups—Vacuum Air Systems—Electrical Systems—The Rockwell Commander 840 System—Hydraulic Systems—Examining a Hydraulic System—Landing Gear Systems—Personal Venture—Flap Systems—Power Steering and Power Brakes — De-Ice and Anti-Ice Systems — Props-Anti-Ice — Windshields-Anti-Ice — Turbine Engines-Anti-Ice — Airframe De-Icing — Environmental Control Systems — Heating Systems — Air-Conditioning Systems — Fire Detection and Extinguishing Systems

4 **Introduction To High-Performance Reciprocating Engines** 64
Geared Vs. Direct-Drive Engines—Don't Get Psyched Out!—The Continental Tiara Series—Modern Turbocharging Systems—Operating a Turbocharger—Oil Systems—Induction Air Systems—Ignition Systems—Fuel Injection Systems—Exhaust-Gas Temperature Indicators—Engine Cooling Considerations—High-Performance Engine Operating Tips—Starting—Takeoff in a Turbocharger—Climb—Cruise—Descent—Landing, Taxi, Shutdown

5 **Flying A Typical Cabin Twin: Piper Navajo** 85
Navajo Preflight—The Navajo Cockpit—Firing Up the Engines—Pre-Taxi Steps—Remember Your CIGAR—Taking Off in the Navajo—Cruising in the Navajo—Landing the Navajo—Abnormal Situations

6 **Flight Planning For Efficiency**...99
General Considerations—Beware of FT Reports—Choosing a Cruise Altitude—Read Your Flight Manual—The Golden Eagle's Flight Manual

7 **High Performance And Solo IFR Techniques** 128
Keeping Your Cool—Keep a Flight Log—Organizing Your Nav Materials—Other Valuable Items—Flying By Yourself

8 **Special Weather Environments** 134
High Level Winds—Icing—Dealing With Ice—Making Your Gear More Effective—Be Prepared—Thunderstorms and Turbulence—Understanding Your Radar—General Rules of Radar—Penetrating a Thunderstorm—Autopilots and Bad Weather—Getting Into the Thick of Things—Approaches in Low Visibility—ILS Below Minimums—Zero-Zero Approach

9 **Pressurization, Oxygen And High Altitude Physiology** 156
The Human System—Maintaining the Proper Environment—Two Types of Oxygen Equipment—Pressurization Systems—Environmental Control System—Operating Pressurization Systems—Pressurization System Emergencies—Air-Source Contamination—Re-Pressurizing at Altitude—Environmental Systems in Pressurized Aircraft

10 **An Introduction To Turboprop Engines** 174
Comparing Turboprops and Gas Turbine Engines—Two Classes of Turboprops—Temperature and Turboprops—Power Controls and Instrumentation—Turboprop Propeller Systems—Starting the Turboprop

11 **Flying A Typical Turboprop: The Beechcraft E90** 190
The Turboprop Preflight—Checking Out the Interior—The Turboprop Fuel System—Engine-Start Sequence—Preparing for Takeoff—Turboprop Takeoff—Turboprop Emergency Procedures—Landing a Turboprop—Dealing with a Dead Battery

12 **Stability Augmentation, Flight Control and Aircraft Monitoring Systems** 206
Understanding SAS—Staying On Top of Your Autopilot—An Autopilot-Controlled Approach—The Flight Director System—Monitoring and Controlling Devices

13 **Long Range Nav And Flight Management Systems (RNAV)** 216
RNAV's Functions—Long Range Nav Aids

14 **Flight Desk Management** 220
Dividing the Duties—Leg-Swapping

Index 223

1 Review of Basic Multi-Engine Techniques

You may be a new multi-engine pilot, with your most recent multi-logbook entry your successful check ride for the rating; or you may have been flying light twins for years; or you may even be a single-engine type with ambitious plans. In any case, a review of the basic techniques of multi-engine flying is time well spent. In the long run, the differences between different types of twin-engine aircraft are more matters of degree than anything else, and a solid grounding in good multi-engine technique will stand in good stead whether you're flying a Beech Duchess or a Convair. In fact, the only twin around that doesn't behave like all the others when one engine suddenly becomes uninterested in further toil (and that's really what multi-engine flying is all about, that sticky moment when suddenly the engines aren't multi anymore) is the redoubtable Cessna Skymaster, or "push-me-pull-you." But since even that ship is available in a pressurized version, would-be pusher-pullers will find considerable information here.

BASIC CONFIGURATIONS

Considerations of the Skymaster aside, let's examine the basic configuration of any twin: a wing, a fuselage, either sitting on top of it or hung underneath it (not that it makes any difference), an engine on each side, and a tail at the back. As long as both engines are running, there's virtually no difference between a twin and a single. If the throttles were taped together, any competent Bonanza pilot could turn in quite a creditable performance in any light twin, or, for that matter, and discounting minor points of system management, any of the cabin-class reciprocating-engine powered twins we'll be discussing later in this book.

No, it's when an engine stops that the multi-engine pilot starts earning his keep. Think back to your training: of all the time you spent in that allegedly multi-engine trainer, how much was really spent with both engines putting out the same amount of power? Not a whole lot, right?

Fig. 1-1. The large vertical stabilizer and rudder on twins is largely required to aid in single-engine operation (courtesy of Cessna).

Controllability and Performance

When we have to deal with a single-engine situation, we're faced with two ultimate problems—that's all. In order of importance, they are controllability, and performance. It's the interaction between the engine(s) out on the wing, and that tail we mentioned earlier, that determines controllability. The power available from the functioning engine, aircraft weight, density altitude, and proper pilot technique determine performance. Note, however, that (at least as I like to teach) performance is subsidiary to control; all the performance in the world will be of little use to you if the aircraft is uncontrollable.

(If I can be allowed a final digression about the Cessna Skymaster, let it be noted that, while its unorthodox design represents definite improvements in the area of single-engine controllability, its single-engine *performance* is no better than that of any other aircraft in its class. Lose one, particularly the rear one, on takeoff in a Skymaster, and while you may find it a bit easier to keep the machine pointed in the right direction, it'll require every bit as much single-engine technique as that of your motors-out-on-the-wing brethren to keep the machine airborne, and it'll be harder for you to identify which engine has failed. There's only one twin flying that requires no special pilot technique to maintain both control *and* adequate performance in a low-level engine-failure situation, and that's Burt Rutan's remarkable Defiant, currently flying only in prototype.)

We'll address the controllability problem first. The situation is a simple one: on one wing, we have an engine providing forward thrust; on the other, a (failed) engine providing rearward drag. The obvious result: a tendency to turn toward the dead engine; just as your car would pull to one side if the brakes on that side were applied suddenly. What means do we have to counteract this yawing tendency? The flight control that we always use to affect movement about the yaw axis—the rudder.

Look at almost any twin, and you'll notice that, compared to a single, it has a disproportionately larger vertical tail and rudder (Fig. 1-1). This isn't just to ensure good directional stability at cruise; the main reason for all that size is to allow the pilot a chance to overcome the yawing moment of a failed engine (Fig. 1-2). Even so, a certain minimum airflow over that big fin and rudder are required to hold the airplane straight against full asymmetric power, and that required minimum gives rise to one of the most important operating parameters for any twin—V_{mc}, or Minimum Control Speed.

The determining factor for this all-important airspeed is a very simple one: it's the lowest possible airspeed at which there's enough "rudder power" available, aided by not more than 5 degrees of bank, to maintain a constant heading with the aircraft in takeoff configuration. The functioning engine will produce maximum rated power, while the *critical* failed engine will produce maximum drag, causing windmilling. Boiled down to simpler terms yet: lose an engine in the air at less than V_{mc} under the above conditions, and there is *no way* the airplane will remain controllable unless airspeed is increased or power on the operating engine is reduced. You can be the most proficient pilot in the world and you will still not be able to maintain control. Violate V_{mc} on takeoff, lose an engine, and *you will lose control of the aircraft.*

In fact, this may be a good point at which to interject a few carefully-chosen words on the subject of twin-engine safety. Intuitively, one as-

Fig. 1-2. T-tails (here on a Beech Super King Air) are becoming increasingly popular for both aerodynamic and styling reasons. Note the dual trim tabs at the elevators' inboard ends. The small horizontal "fin" just below the American flag is a type of VOR antenna. This aircraft is equipped with pneumatic de-ice boots for the stabilizer.

sumes that twins are safer than singles. They're not. In fact, let's deal with the bleak statistics: the fatality record of twins is considerably higher than that of singles. Why is this?

The answer comes right back to our two salient points: controllability and performance. True, the situation is mitigated when we consider engine failures during cruise at altitude. The pilot of the single has no recourse but to make a forced landing, while the twin driver can generally continue to some airport, if not necessarily the destination, for an uneventful landing.

Engine Failure At Takeoff

Consider the worst-case failure, generally conceded to be an engine failure immediately after takeoff, and the situation changes. The pilot of the single-engine aircraft has no recourse but to land, usually straight ahead. In an impressive number of cases, the outcome of such a maneuver is entirely successful from the passengers' standpoint and possibly from that of the aircraft or its owner.

Theoretically, the pilot of a properly-flown twin-engine aircraft is in a better situation. Confronted with an engine failure at liftoff, he or she should be able to stay in the air, get back around, and land. This is true *if, and only if*, the remaining engine is performing up to specs, the takeoff has been properly planned and calculated *and the pilot technique is virtually flawless*. Sure, the aircraft specifications indicate, or should indicate, the distance required to clear the proverbial 50-foot obstacle after losing an engine on takeoff. Remember, though, that those figures were arrived at by using a brand-new airplane, flown by an experienced test pilot who was really sharp on single-engine technique, and who *expected* that engine to fail when it did. There's little or no margin for error in such a situation, and what little margin there may be can be eaten away very quickly by the entirely natural reactions of shock and surprise that may accompany a real engine failure. Inevitably, pilot performance deteriorates under these conditions. The first casualty is generally that level of technique required to extract the airplane's maximum performance, while in more extreme cases, controllability may disappear, as well.

Frankly, until you get into turboprops, the single-engine performance of medium and heavy twins isn't exactly sparkling anyway. My old instructor used to describe the remaining functional engine of one type we flew as "just powerful enough to get you to the scene of the crash." It would appear from accident reports that, at least in some instances, the prudent course after losing one engine on takeoff in a twin would have been to reduce power on the remaining engine and land straight ahead, single-engine style. In some cases, it was the pilot's frantic efforts to avoid this course which resulted in a far worse loss of control, with much more serious consequences than a straight-ahead landing might have had.

The concept of V_{mc} then, is meant to protect the pilot and passengers from the consequences of an engine loss at low speed and low altitude, and, properly applied, it does just that. Remember, though twins usually have higher takeoff and landing speeds than singles, so that if something does go wrong, it'll do so faster, and any impact with the ground will tend to be more severe. Remember, too, that the twin-engine aircraft is statistically twice as likely to lose an engine on takeoff than a single.

We'll have more to say on safety and professionalism later in the book. For the moment, note the previous points and realize that we mustn't be lulled into a feeling of invulnerability just because there are two motors out there.

I feel that in many ways V_{mc} is an airspeed of considerably more importance to the multiengine pilot than the aircraft's stalling speed. The airplanes we're discussing in this book will seldom, if ever, approach the stall except in emergencies or training situations. On the other hand, they'll be exposed to speeds near and/or below V_{mc} on virtually every takeoff and landing. We'll discuss V_{mc} at greater length later in the book. For the moment, we'll close our considerations of aircraft controllability with the thought that as long as we're flying above V_{mc}, the airplane will remain controllable (with proper pilot technique, of course) through the worst-case engine failure. Seldom, if ever, is there justification to be airborne at or below V_{mc} in almost any of the aircraft we'll be discussing.

Dealing With Engine Failure

If aircraft controllability is a fairly simple and straightforward subject, then single-engine performance is a somewhat more complex one. Instead of one clear-cut speed, we now have several to deal with, and our choice of which one is appropriate may vary with the situation. Again, the worst case is generally deemed to be an engine failure on takeoff (although, personally, I think that losing one on a go-around can be worse yet). The speeds we have to be aware of, starting with the lower one, are: the airspeed for best *angle* of climb (V_{xse}), and, slightly higher, that for best *rate* of climb (V_{yse}). These speeds, and the aircraft climb performance associated with them, depend of other factors, such as aircraft configuration, making the performance picture a bit less clear-cut than the controllability aspect.

Rather than engaging in a rigorous discussion of all the factors, let's proceed empirically and try to figure out what it takes in the cockpit to lose an engine and get away with it.

What we're really confronted with is a situation which can range from "ho-hum", a gradual power loss at cruise, to a real "Oh-my-god" takeoff. Since the latter situation can happen so fast that there's no time available to think things over, what you really need is a set of habit patterns, conditioned responses, that are completely automatic. I don't mean that you have to go through the complete routine of securing an engine with lightninglike speed and robot precision, but there *are* some critical actions that are required *right away* in an actual engine-failure situation. These shouldn't be memory items, they should be ingrained, a real gut-level reaction. While these conditioned reactions won't deal with every aspect of the situation, they'll buy you and your passengers enough time to get things squared away.

Fly The Airplane

What are the priorities, then? The first one is almost ridiculously simple, but the files show that a sad number of pilots forget it. They get so engrossed in things like identifying and securing the bad powerplant that they just ignore the need to *fly the airplane!*

What do I mean by that? Nothing more than retain control of the situation—apply rudder to hold the heading, and start the airspeed toward V_{xse} or V_{yse} as appropriate. If the gear's already up, or coming up, try to keep the airplane from settling (but don't get below V_{mc}!). If it's still down, and it should be until there's not enough runway left to land on, consider pulling power off the remaining engine and landing. Even if you go off the far end of the runway, it's likely to be at very low speed, and damage will be minimal. If you have any doubt that you can continue the takeoff safely, put her down!

You'll notice that I said to *start* the airspeed toward the required value. What I mean here is that you have better things to watch at this point than the airspeed indicator, at least for the first second or so. Again, the main thing you have to do is fly the airplane. When I instruct multi students, I try to introduce the concept of airspeed *trends* as well as actual speeds. In other words, the pilot should not only know what the ASI might read at any given instant, but he should also recognize which way it's moving. In general, the trend of airspeed immediately after an engine failure is to *decrease*; thus, it's obviously good practice on any takeoff to get *past* V_{yse} as soon as safely feasible. That way, if there is an engine failure, the airspeed will tend to start in the right direction by itself. You can see that it's good practice to reduce your exposure to speeds below V_{yse}, the blue radial line on the ASI, to a minimum. By the same token, a number of manufacturers have taken to the laudable practice of publishing a suggested "safe single engine speed" (V_{sse}), generally slightly higher than V_{mc}, which provides a very valuable guideline for normal operations.

Very well: you've just lifted off, an engine has failed, and you're flying the airplane for all you're worth. You're not out of the woods yet, but you've bought yourself a couple of seconds. What's the next thing you need? You need all the *power* that the good engine can produce, within its limits, of course.

In most of the recips and even some of the turboprops with automatic engine limiters, then, the next step is a simple one—move all the handles on the power quadrant as far forward as they'll go. When you're training, you'll probably be drilled on the time-honored sequence of mixture, prop, and throttle, and that's probably what you did on your check ride, too. Very nice, but forget it (again, we're talking about recips at this point). It's elegant, but it takes time. Instead, put the flat of your hand across the back of all six handles and just *shove*!

Clean Up The Airplane

What effect will this have? Hopefully, none at all. In a normal takeoff in almost every recip, all the handles will *already* be all the way forward. That's fine, but in case one of them has crept back, you've now ensured that you've got all the power available from the good engine. In actual practice, you'll fly the airplane and add power virtually simultaneously. Now you've bought even more time for the next step—*clean up the airplane.*

Perhaps a better term than "cleaning up the airplane" might be "reconfiguring for maximum climb performance." While we now are well on our way toward having the situation in hand, having maintained control of the airplane and added power, the airplane's *configuration* is still far from ideal for continued flight with one engine. The landing gear, and perhaps the takeoff flaps, are still extended, and we're using a lot of rudder, with its attendant trim drag, to fight the even greater drag of the failed, and now windmilling, engine.

Here, again, there can be no absolute rules, because we're now getting into an area of considerable variation between aircraft types. In some, the landing gear has more drag than the flaps; in others, the situation is reversed. Moreover, some types depend sufficiently on the extra lift of takeoff flaps so that retracting them too soon after liftoff with engine failure virtually guarantees settling back to the runway.

One good rule of thumb is fairly obvious: if we can maintain altitude, or, better yet, a positive rate of climb, the landing gear can be retracted. A few ships have a tendency to settle slightly during gear retraction, due to the additional drag of opened gear doors, but this effect is generally quite small, and the increased performance once the wheels are safely indoors makes it worthwhile.

Similarly, if the airspeed tendency is toward an increase, retracting the flaps (at least on most aircraft) will further increase climb rate and angle.

We've now reached the point at which we can draw a deep breath and, paradoxical though it may seem, take our time considering our next move. Unless we're grossly overloaded or departing from a runway on which we had no business in the first place, the airplane's in a position where it will continue to fly for a few seconds, despite the massive drag of the windmilling propeller. Obviously, the sooner we identify and feather the dead engine, the better off we'll be, up to a point. Sure, popping an engine into the feathered position just as soon as you've maintained control and added power, snapped up the gear and flaps, is impressive, and it *does* reduce drag drastically—*if*, that is, it's the correct engine! The situation just isn't going to deteriorate that quickly during the time it takes to calmly and deliberately identify and *verify* the failed engine, but it will deteriorate quickly indeed if you rush to feather what turns out to be the only good engine you have.

"Dead Foot, Dead Engine"

The classic rule for identifying a failed engine, "dead foot, dead engine", holds as true now as ever, but I'd like to add one caution—it only works for sure if you have the airplane well under control, with the wings level. I've actually seen pilots holding so much aileron against a dead engine that they had to hold rudder *into* it. If they had followed the "dead foot" rule, it might have become true in its most macabre meaning! On the other hand, if you've got the wings level, and are close to the ground, there's no reason not to use the classic rule. I've personally gotten into the habit of tapping the "dead" foot against the rudder pedal a couple of times, sort of like a circus horse telling you its age. There's no way you can tape the "live" foot, since as soon as you try to lift it, the rudder pedal will come back with it.

With the dead engine identified, the next step is to *verify* it. Put your hand on the appropriate throttle and *look at it* to make sure it's the right one, and slowly bring it back toward idle. If the airplane's yawing tendency gets no worse, you're on the right track; if it suddenly improves, be careful, you're probably pulling power on the good engine, and you will lose altitude. The moral is: take your time, make the throttle movements nice and slow. You may even find that the "dead" engine, while definitely not up to par, may still be putting out some useable power. Whether or not you should feather it under those circumstances is a complex decision based on many factors including current performance and your experience and knowledge of the particular aircraft type.

Feather the Engine

With the dead engine verified, we can finally go ahead and *feather*, and the improvement will be dramatic. Again, take your time, look at your hand to make sure you have the correct prop control, and if you have a copilot, announce which engine you're about to feather in a loud voice. Then give him an instant to scream. "No, that's the wrong one!!" before you finally move the propeller control through the detent and into the feathered position.

Whew! We made it! The airplane's flying, the configuration is correct, and the dead engine is feathered. With each step we've bought a bit more time, reduced the pressure on us and added a little margin for error. By now, we've arrived at the stage where we can consider relatively calmly what airspeed is appropriate—V_{xsw} to clear obstacles, or V_{yse} to gain altitude as quickly as possible. We also have time to decide whether to secure the dead engine (prepare for protracted single-engine flight) or attempt to determine the cause of stoppage, and restart it.

Gaining as much altitude as possible in the least amount of time is always a wise course in an emergency. In fact, this holds true from the instant of engine failure onward. While an airplane tends to decrease speed fairly slow after losing an engine, since it has considerable forward momentum, it tends to lose altitude rather quickly since gravity is dragging it downward. Furthermore, while we can reduce drag fairly soon by properly reconfiguring the aircraft and feathering the dead engine, there's nothing we can do about gravity—it's always there. Thus, even during a multiengine takeoff without an engine failure, it's good practice to climb at or near the best multiengine rate of climb airspeed until you have gained at least some 500 feet of altitude. I like to teach my multiengine students to remain at full takeoff power, with no reductions, until reaching that altitude. With takeoff power, those first 500 feet are gained so quickly that there's no undue strain on the engines, and statistics show that the first power reduction is a point at which a disproportionate amount of takeoff engine failures occur. In fact, that 500-foot cushion goes far toward guaranteeing a safe return to land after an engine failure, even in the face of less-than-brilliant pilot techniques. It makes the conditions of an engine failure similar to one at cruise rather than a critical one near the ground.

Critical Engine

Speaking of "critical," you'll recall that I promised to touch on the idea of the "critical" engine. This is merely the FAA's way of saying the engine that you'll miss the most if it fails is the one whose loss most adversely affects the airplane's performance. In twins with props that turn clockwise as seen from the rear, it would be the left engine. At high angles of attack, as found during takeoffs, the downgoing (right-hand side of each engine) propeller blade produces somewhat more thrust than the upgoing one, so the airplane responds as if the engine were a bit further to the right than it actually is. The result of this is that a still-running *left* engine acts as though it were closer to the fuselage than it really is, thus reducing its leverage and requiring less rudder to hold it than a still-running *right* engine, which acts as though it were further outboard than is actually the case.

Fig. 1-3. Classroom training can be an invaluable aid to checkout in new and complex aircraft, and it may bring a reduced insurance premium, too!

Even the Cessna Skymaster has its critical engine, since it flies better on the rear one alone than on the front.

Here's an area in which knowledge of your particular aircraft can be a valuable asset (Fig. 1-3). While the FAA definition of a critical engine is a valid one, it fails to take into account certain operational considerations. Consider, for example, an aircraft with only one hydraulic pump, and that one on the right engine. Lose that engine, and while the airplane may fly marginally better on the left one than on the right, you are faced with the problem of having to use the hand pump to retract gear and flaps, which is a time-consuming process at best, This is the last thing you need to bother with at this point. Most modern aircraft have tried to eliminate such pitfalls, but the pilot should be aware of the possibilities and have a plan of action filed away for each emergency situation.

Basic Reactions To Engine Failure

Let's briefly reiterate the steps that have to become virtually instinctive reactions during an engine failure:

—Fly the airplane
—Add power
—Check proper configuration
—Identify the failed engine
—Verify your identification
—Feather
—Secure or restart the failed engine

Whether you lose an engine on takeoff, during a go-around or at cruise, these stops remain equally valid. If you have an automatic reaction to an engine failure, about the worst thing that can happen is that you'll overreact if one engine quits at cruise. You'll start slowing toward V_{yse}, apply full power and check your already-retracted gear and flaps (noisy and embarassing, but safe.) More often, of course, a little inner voice will tell you, "It's all right, take your time, we're in a cruise situation." However, it's good to have the automatic reactions ready, and then suppress them if necessary. The larger and faster the airplane you're flying, the less time and margin for errors you'll have when the chips are really down.

2

Handling Large, Complex Airplanes

In Chapter 1, we looked back at a couple of the areas covered in basic multi-engine training. Obviously, there's more to flying a twin than dealing with an engine failure at takeoff. I chose that particular area for in-depth discussion partly because it's one of the most critical, and partly because it embodies most of the skills and habit patterns required to handle almost any multi-engine emergency, no matter what airplane you may be flying.

Now, we'll examine some of the factors that make flying larger, heavier aircraft different from handling the light training types in which most multi-engine instruction is given (Fig. 2-1).

Light Twins vs. the Large Aircraft

As I've already mentioned, the difference between flying light twins and flying the medium and heavy types is more a matter of degree than of actual changes in technique. As a matter of fact, one can almost look at it as a one-day exchange—you can use large-aircraft techniques to handle even the lightest of four-place twins, and they'll work just fine. You'll actually benefit by an increase in smoothness and comfort, a benefit that will of course be shared by your passengers. On the other hand, the reverse doesn't necessarily hold true—try to use light-twin habit patterns and responses to handle one of the larger aircraft, and your performance will be uncomfortable at best, and unsafe at worst.

Well, what are the differences, then? After all, any aircraft has to adhere to the same set of physical laws, doesn't it?

Indeed it does, and that's what gives rise to the required differences in technique. The old addage, "The bigger they come, the harder they fall," doesn't hold true here. But, what does apply is the area of physics that deals with inertia, momentum, and kinetic energy. Reducing this to the simplest terms, we'll find that the larger and faster-moving an airplane we're dealing with, the more energy and time will be required to change its

Fig. 2-1. The Mitsubishi Solitaire is typical of the turboprop corporate twins the professional pilot may expect to fly.

speed and direction, or both. Remember Newton: "A body in straight motion will tend to retain that motion unless disturbed by some outside force."

From a practical standpoint, this means that it's much more important for the pilot of a large aircraft to always remain well "ahead of the airplane." There are areas in which the larger airplane is more forgiving than the small one, but this isn't one of them. A larger airplane always needs more room and more time for any maneuver, simply because we're

Fig. 2-2. The Piper Cheyenne III is fitted with the P&W PT6A and has a single-engine service ceiling of 12,500 ft. (courtesy of Piper).

Fig. 2-3. Cheyenne III instrument panel (courtesy of Piper).

dealing with a larger mass, moving at a higher rate of speed, as compared to the lighter aircraft.

Speed plays an important role, also when we consider that the kinetic energy of any moving object is determined by its mass times the square of its speed.

Thus, for example, we can compare a light four-place twin and a medium pressurized one on final approach (Figs. 2-2, 2-3 and 2-4). If we assume that the light aircraft has a landing weight of 4000 lbs. and the heavier one a landing weight of 7000 lbs., and final approach speeds of 85 and 110 knots respectively, we find that the medium ship has almost *three times* the kinetic energy of the light one. This doesn't mean that it'll take exactly three times as long to effect any change in its direction, since it has proportionately larger and more powerful controls, but it *will* require considerably more room and time for maneuvering.

What this means for you as a pilot is a factor that can work both ways—on the one hand, a larger aircraft is inherently smoother, especially in turbulence. Also, its very size and mass make it want to "keep going the way it's pointed." On the other hand, this same tendency can make things difficult if you have to make a sudden change in the flight path, whether horizontal (turns) or vertical (climbs or descents). The solution is a simple one, and one that will contribute greatly to passenger comfort as well as safety—simply avoid situations in which sudden changes in the flight path are ever necessary!

Fig. 2-4. Typical interior of the Cheyenne III (courtesy of Piper).

Of course, a lot of this will come to you more or less instinctively. When you make the transition into a heavier aircraft, your own feeling for its responses to control pressures and movements can go far toward telling you how the airplane will respond, and how it "wants" to be flown. This only works, of course, if you have some feeling for what you're doing. While it cannot be disputed that larger aircraft tends to be flown more by instruments, and "by the book," than their smaller brethren, a certain amount of "seat of the pants" feeling remains a very valuable asset to possess. Moreover, the time may come when you'll encounter some situation that will demand enough of your attention outside the airplane that you won't have time to scrutinize the instruments and fly mechanically. So, as you fly, keep in mind that there's more to it than being a human autopilot.

Bigger Planes Give Smoother Rides

In a cruise situation, of course, there's seldom a need for sudden changes. However, even here you'll find certain differences between the handling of larger and smaller planes. On the whole, larger airplanes are less susceptible to disturbance by turbulence, so you may find that holding your heading and altitude are actually easier in larger aircraft. If your heading does wander off, though you may find that it'll take a bit more effort to get back in the groove than in the nimble smaller aircraft. The higher aircraft speed will tend to exaggerate the effects of even small pitch changes; while the aircraft itself may have less tendency to "hunt" in altitude. Any pilot-induced pitch changes will show up at once as surprisingly high rates of climb or descent. The actual pitch change may be nearly imperceptible to you at first. One of the most common experiences for the tyro high-performance pilot is for him to sit at the controls, congratulating himself on the smoothness of his performance, only to glance at the panel and realize that he's smoothly climbing or descending at up to 1000 fpm! By the same token, given the greater mass and speed of the airplane, once you've established one of these inadvertent altitude excursions, it'll take more airspace than you're used to before you get it under control.

The higher airspeeds at cruise also make themselves known in a rather less obvious fashion, one that becomes more evident the faster you're cruising—it may take what you initially considered remarkable angles of bank to make a turn. This is because increased airspeeds leads to *reduced* rates of turn for a given angle of bank. If you need to make a standard-rate turn, you may find yourself banked quite steeply before the turn needle lines up under the little doghouse! At the 150-knot cruise speed of a typical light twin, for example, a standard-rate turn requires 22 degrees of bank. At the 210-knot cruise of a pressurized twin or slower turboprop, the same turn rate requires 30 degrees. For faster turn rates, the bank goes up quite quickly. If you need twice-standard rate, as might be the case if you're making a too-sharp intercept to a VOT course or localizer, you'd need over 50 degrees of bank in the faster airplane (a prospect not likely to be viewed with delight by the passengers as they strive to hold onto their drinks under the roughly 1.6 g-load).

Stay Ahead of the Game

Of course, it's seldom necessary to do much maneuvering at cruise speed. Under normal circumstances, the occasional course or altitude change is about all that will come up. Approach and landing situations, though, often require considerable modification of the flight path, and this is where the larger airplanes differ the most from their smaller counterparts.

At the lower airspeeds used for approaches and pattern work, the pitch sensitivity and large bank angles mentioned above are less of a problem. Getting to those lower airspeeds, though, can require considerable forethought, since larger aircraft are not only heavy, but generally quite clean aerodynamically as well. Couple these characteristics with initially high cruise and descent speeds, as well as the need to avoid sudden engine cooling (which will be covered more completely in a later chapter), and it becomes evident that these aircraft need considerable room simply to slow down (often several miles). This is an area where learning to stay ahead of the aircraft becomes very important.

Another point to be made here is that larger aircraft are often more sensitive to configuration changes than smaller ones. That is, for example, a light twin might well be able to initiate and complete a go-around in full landing configuration, while a larger aircraft would require reconfiguration for the successful completion of the maneuver. The wider the range between cruise and approach speeds, the more likely this condition becomes, and the fewer the options available to the pilot in final approach configuration.

Learn to Commit Yourself

As a result, the pilot of the medium or heavy twin must get used to the concept of commitment. By this, I don't mean that he or she has to be dedicated to a career in aviation (although that doesn't hurt), I simply mean that there are various times in each flight where the pilot must realize that the aircraft is more or less irrevocably committed to a particular course of action and that from that point on, any attempt to change that course of action might not be successful.

It's obvious that takeoff and landing situations are the areas in which these commitments appear, and we'll examine each in turn. Some areas are, of course, pretty clear-cut. For example, should an engine fail at some speed less than V_{mc}, there's only one possible course of action—chop the power on the remaining engine and stop, hopefully on the runway.

The first gray area, then, appears as soon as V_{mc} is reached, but possibly still a knot or two short of liftoff speed. This is when you must decide to continue the takeoff, or abort. Generally, if it's at all possible to stay on the ground, you should. If the runway is so short that you can't accelerate to V_{mc}, lose an engine, and still stop comfortably short of the end, then you and your airplane don't have any business on that airfield.

This gray area, or "area of decision," actually extends from the point at which V_{mc} or V_r (rotation speed) is reached to the point at which the airplane is too high or too fast to be brought to a stop on the runway. At the far end of the area of decision is another "commitment point." Just as the pilot is committed to abort if an engine fails below V_{mc}, the pilot is committed to continue once he has reached the point where there's no way he could get back down and stopped in time. Many avoidable accidents have been caused by last-minute changes and violation of such commitments. When flying any twin, but particularly the heavier ones, it's of paramount importance to make and respect the appropriate commitments for each and every takeoff and landing.

The choice of takeoff commitment point depends on many factors, including aircraft loading and performance, runway length, terrain off the end of the airport, and so forth. Pilot knowledge and experience in the particular aircraft are important, too. Obviously, the longer the runway, the longer one can delay being committed to continue a takeoff and fly. Making takeoff commitments has become somewhat easier since manufacturers have started listing not only the accelerate-stop distance, but also the accelerate-go distance. (The distance required to accelerate to liftoff speed, lose an engine, and continue the takeoff to an altitude of 50 feet.) What I cannot stress strongly enough is the importance of being aware of these commitments, and *sticking with them.* At most decent-size airfields, runway length and aircraft climb capability combine to permit the laudable practice of leaving the landing gear extended until the aircraft is committed to further flight. This not only lengthens the zone of decisions, that is, the zone in which the pilot still has the option to abort or continue, but palpably underscores the commitment to flight once that point is reached and the gear retracted. While it may seem to some that lengthening the area of decision increases the uncertainty factor of a takeoff, I maintain that the more options you have, the safer you are, so why rush to give up one of your two takeoff options before you have to?

By the same token, there comes a point on each landing at which the airplane is committed to land. Here, the primary consideration is the aicraft's performance in the event of a single-engine go-around, and this is an area in which performance is generally quite poor. In fact, I personally feel that a single-engine go-around is the ultimate test of an airplane's single-engine capabilities.

Forced to Go Aground

Let's look first at the less-severe case: an aircraft forced to go around after approaching with an engine already secured. You'll recall from your multi-engine training that you were warned not to make the transition to the full, high-drag landing configuration until a landing was assured, that is, until you were *committed to land*. This is sound practice, of course, but it can lead to a rather paradoxical situation—you can actually descend with one engine feathered to an altitude lower than that at which you would be committed to land if both engines were running but one were to fail!

Why is this? Because initiating a two-engine go-around and having an engine fail at exactly that instant is the worst possible case for an engine failure, even worse than losing one on takeoff. It is only the fact that you're already (or still) well above the surface that mitigates it somewhat.

Lose an engine on takeoff, and, while the situation is a tricky one, you still have a couple of things in your favor. The operating engine is already at full power, and the aircraft is in a takeoff and initial climb configuration, with an appropriate flap setting. Moreover, it's at least beginning to climb. Now compare that with an engine failure on final—the airplane is as dirty as possible, with full flap; the other engine is at low or idle power, and may require several seconds before delivering anything useful. The airspeed is already low, and will decay even further with all that drag from the flaps, and there's a *significant downward component to the flight path*. Remember, larger and heavier aircraft need more time and space to change direction and velocity, and that certainly holds true here. It's not at all difficult to use up a good (or bad) 500 feet transitioning to a solid climb under such circumstances, so you could consider that a reasonable "committed to land" altitude. On *any* multiengine approach, then, you'd have to have that below 500; I'm committed to land. It *does* mean, however, that you have to thing, "I'm below 500; I'm committed to land *if I lose one engine now.*" (The 500-foot figure may vary from airplane to airplane, of course, as well as depending on loading and density altitude.)

Where the paradox comes in is the fact that you can shave a bit off that margin if you're already approaching with an engine caged. Under such circumstances, it's unlikely that you'd be using more than the minimum flap required, and you might well be holding off on the gear and flaps until you were at, say, 250 feet, ½ mile out on final approach. Thus, you could go around from that point without too much trouble; at any rate, you'd be better off than if you'd lost one engine on takeoff. Again, though there's a commitment to be made—with a single engine, once the gear and flaps are down, you're committed to land. If the proverbial truck pulls out on the runway in front of you, you're less likely to damage your airplane landing in the grass next to the runway than by trying to clean the airplane up again and go around on one engine.

Handling Techniques

While the major differences between larger and smaller aircraft handling techniques occur in the takeoff and landing phases, there are a couple of points to be made about cruise and letdown planning and handling as well. If you've gotten used to the performance of the smaller twins, you

Fig. 2-5. Cessna Titan has single-engine service ceiling of 10,100 ft; cruises at 217 kts at 20,000 (courtesy of Cessna).

may be a bit surprised at how quickly things happen in navigation situations. Intersections are reached almost as soon as the needle becomes "alive," and course intercepts require considerable "lead" unless high-g pylon turns on instruments are your idea of a good time. Even simple course changes over a VOR or waypoint require a bit of forethought, especially if the turn is a relatively larger one. Remember, the airway is only 4 nm wide over most VOR's. If you come steaming up the VOR at some three or four miles per minute and wait for the flag to change from "to" to "from" before starting your turn, you'll very likely end up well outside the protected airspace of the airway.

Descents are an area of particular interest, and the factors that dictate descent planning can vary considerably from aircraft to aircraft. Here's an area in which the turboprops are actually a bit simpler to handle than the big recips, since it's relatively easy for them to remain at high, efficient altitudes until fairly close to their destination. Then they can descend steeply and rapidly with little concern for rapid engine cooling or the need to maintain sufficient power for cabin pressurization. The recips, on the other hand, are constrained to rather shallower descents, since power must be kept high enough to prevent chilling the engines and maintain enough turbocharger output to keep the cabin "pumped up."

In either case, the aircraft are likely to be moving at a pretty good clip when leveling off at a lower altitude, and here's where poor planning will really show. A heavy, clean machine is naturally reluctant to shed airspeed very fast, and it can be more than embarassing to arrive over, say, an outer marker at exactly the right altitude, but 30 or 40 knots above maximum gear extension speed! True, the more modern ships try to fit more efficiently into the ATC environment by offering high gear and flap extension speeds, but planning is still important. At first, you may find the large numbers a bit disconcerting. For example, if you're cruising in a pressurized recip at FL 210, and plan a landing at a sea-level field in VFR conditions we have to lose 20,000 feet to get down to pattern altitude. An airplane rate of descent of 1000 fpm will give us a comfortable cabin descent to prevent "ear bumps," and at a decently warm descent power setting, we might be making good a ground-speed of 240 knots—four miles a minute, so we'd have to start down a good 80 nm out. In fact, we should tack on 5 nm or so in order to slow down so at about 100 nm out we might want to begin negotiating with the ATC folks for a descent.

You can see that DME is very handy in such operations. In fact, once you've figured out how much time it'll take to get down, distance becomes a secondary factor, and if your DME is one that displays time to station as well as ground speed, it's easy to keep running tabs on your descent, adjusting the rate as necessary. Even handier is a sophisticated RNAV with VNAV functions. Such systems will be discussed at greater length elsewhere in this book.

Weight And Balance

Let's cover one final area of operational difference before moving on to a discussion of aircraft systems—weight and balance.

There's nothing really different or mysterious about doing the weight and balance work on bigger airplanes than you're used to; even the arcane calculations you'll find in ATP-course workbooks are based on the same old weight and moment arm principles you learned to know and loathe when you were studying for your Private Pilot's license. The only change in bigger airplanes is that the weights are larger, the moment arms are longer, and that there are generally more places you can stash your disposable loads of fuel, passengers, and bags.

Loading Flexibility

Many large aircraft are designed with a certain amount of "loading flexibility" (Figs. 2-5, 2-6 and 2-7). This is the manufacturer's polite way of saying that if you fill all the tanks, all the seats, and all the baggage holes, the thing won't fly. A more reasonable way of looking at the situation is to

Fig. 2-6. Titan instrument panel (courtesy of Cessna).

Fig. 2-7. Cessna Titan cabin (courtesy of Cessna).

assume that you're dealing with a "basic airplane" that fits the particular requirements of your own most common mission, but whose performance can be extended in various directions at some sacrifice in other areas. For example, your typical mission might be to fly four people 500 miles with reserves and ample baggage, but in the same airplane you could fly six people 300 miles, or one key engineer and his tool box 900 miles, or a badly-needed 350-lb blivet assembly to a factory 700 miles away. In a way, loading flexibility works in favor of the aircraft buyer, since you end up paying primarily for the performance you need all the time, rather than tying up capital in more airplane than you need most of the time.

For the pilot, the loading flexibility of a larger aircraft means a bit more preflight attention to what goes where. Often, large-aircraft baggage bays are proportionally further from the center of gravity than those in smaller ships, making use of empty space in tailcones and pointy noses, so it behooves you to be well aware of where those golf bags, tool kits, and portable typewriters are going! Many corporate pilots have gotten into the habit of carrying a little spring scale and cargo net arrangement, sometimes sold as "de-liars" at fishing stores, for more accurate rampside baggage judging than the old "heft and strain" method.

Fuel Loads

Fuel loads, too, can be a bit more critical in the big airplanes. Gone are the days when you could just pull up to the pumps and say, "Top 'er off." (Of course, quite apart from questions of aircraft performance, few people today can *afford* to "top 'er off!") No, you're now in an environment in which you, the pilot, will have to monitor how much fuel goes into which tank, whether by checking tabs or slots in the fuel filler, or by watching the meter on the fuel truck, it depends on the aircraft. Moreover, the trend at certain airports in recent years has been to hire only some of the barely domesticated types of primate as line crew, so you may have to keep a close eye on refueling operations to make sure that the right amount goes into the right tank, or that it's the right kind of fuel in the first place!

We'll discover the idiosyncracies of some kinds of fuel systems in more detail in the following chapters. Let's close with the final note that larger airplanes use more fuel than smaller ones, and you're now entering an environment in which a fairly accurate knowledge of aircraft gross weight is necessary at all times. In light twins, you've probably gotten used to using the same basic operating speeds for every takeoff and landing. No more, my friend, from now on, every takeoff should be calculated by the book, and the data entered on a card or form easily visible in the cockpit. Moreover, and here's where the higher fuel consumption plays a part, your gross weight may vary quite widely from landing to landing, and you can no longer use just one approach speed every time. It doesn't take long figure out your landing weight and calculate the proper approach speed for it, and you may be surprised at how much runway you've been wasting in the past by using only one speed. Such calculations are normally part of your preflight and flight planning, but you should be able to figure out a new landing weight and approach speed inflight in case you have to divert to a new destination.

Relax—It's Not That Complicated

This whole section may make it sound as if flying a medium or heavy twin requires as much paperwork as flying an airliner. Relax—that's not quite the case. The added workload shouldn't amount to more than five minutes at preflight (less if you have a good calculator), and it'll add a great deal to the precision and professionalism of your operation, not to mention your confidence and peace of mind. Let's face it, today's advanced corporate twins are a lot closer to airliners, in performance and sophistication, than they are to light twins, and they should be flown as such.

3 Large Aircraft And Their Systems: Complexity vs. Redundancy

So far, we've been talking about the operational aspects of flying larger aircraft. Now, let's spend some time looking at some of the physical features of the aircraft themselves—their on-board systems.

Of course, any aircraft has its systems: fuel, electrical, heating and ventilating, and so forth. As we get into larger aircraft, though, these systems became more complex, and are sometimes augmented by others, as well. Sometimes the complexity is needless, a mark of poor design; at other times, though, a more complex system offers more redundancy, more ways of dealing with individual component failures. To use a popular engineering buzzword, a more complex system can be, if properly designed and operated by someone who knows what's what, more "failure-tolerant."

Fuel Systems

We'll start with fuel systems. The basic elements of any such system are simple: someplace to put the fuel, some way of getting it into powerplants(s) and controlling its flow, and some way of figuring how much of the stuff we have. The ultimately simple system, for example, is one like that found in the Piper J-3 "Cub." This aircraft has a simple tank mounted higher than the engine for gravity feed, a copper tubing fuel line to the carburetor with a valve to shut it off, and a cork float to tell the pilot how much fuel remains.

Unfortunately, it isn't that simple for most larger twins. The requirement to carry enough fuel for longer ranges necessitates installing tanks wherever there is room for it in the airframe without causing insurmountable CG shifts with fuel burnoff. As soon as we have multiple tanks and multiple engines, we need various valves and interconnections to tie the system together, multiple means of measuring fuel quantity in the various tanks, and specific operating procedures for proper fuel management. We're also faced with the prospect of pilot error and improper

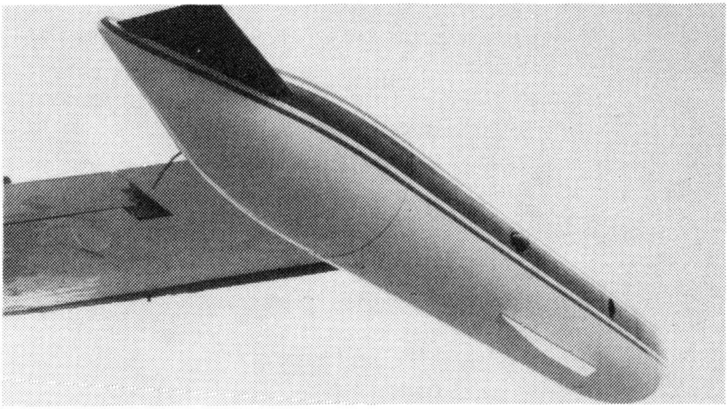

Fig. 3-1. Some aircraft, like this Mitsubishi MU-2, use tip tanks to extend fuel range. Note the stabilizing fin at the rear, and the strake on the bottom of the 90-gallon unit.

operation; while the best-engine systems go far toward preventing this, others virtually invite it.

The ideal system, of course, would be an outgrowth of the Piper Cub: a single immense tank, two copper fuel lines (one to each engine), and a cork float. Any fuel aboard the airplane would be available to either (or both) engine(s). Some of the earlier Aero Commanders approached this enviable ideal, and the high wing even allowed for gravity feed! The fuel systems of some more recent aircraft, notably the Aerostars show an attempt toward a smaller system, with reduction of pilot workload to a minimum, but this hasn't always been successful. The Aerostar, for example, fulfilled its function when operated properly, but a series of fuel mismanagement accidents revealed that certain types of improper operation could cause a dual engine failure even though fuel remained available in one tank!

Systems of the minimum-fuel-management type generally have a large center-section of fuselage tank from which both engines draw fuel in normal operation. Additional fuel carried in outboard or wing tip tanks (Fig. 3-1 and 3-2) is fed directly to the engine on that side of the aircraft in some installations (Aerostar); in others, it's either available at a special fuel selector position (older Commander) or transferred by pumps into the center tank.

When such a system is working properly, it is a delight to use, since using it requires no further fuel management than ensuring that there's adequate fuel aboard in the first place and that each engine's individual fuel shutoff valve is open. (Why have individual shutoff valves at all? I hear you cry. Because you need some way of isolating fuel at the firewall in case of engine/nacelle fire, that's why.) If there are no malfunctions, any fuel aboard the machine will find its way to the engines sooner or later. Even here, though, there can be some pitfalls. For example, some systems are

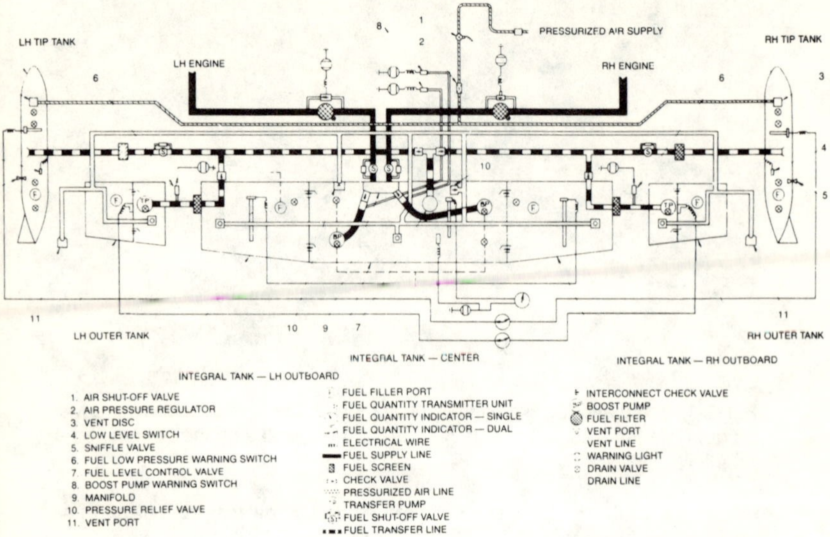

Fig. 3-2. In this MU-2 system, fuel is transferred from the outer tanks to the center integral tanks by electrical transfer pumps. The tip tanks are pressurized with engine bleed air, forcing their fuel into the transfer lines; low-level switches turn off the air as the tip tanks empty, avoiding air in the fuel lines. Air pressure is available to both tip tanks even with only one engine running.

arranged such that the center tank is the only one which can feed both engines, while the outboards are accessible only to the engine on their respective side of the airplane. Result? If there is an engine failure, any fuel in the outboard tank on that side might as well be on the moon as far as single-engine operations are concerned. If you're flying such an airplane, a wise course of action would be to burn off the outboards as early in the flight as is feasible. The longer you keep fuel in the center tank, the more you will have available for single-engine operations.

Almost as simple, and in some ways more versatile, are some of the systems appearing in more recent aircraft, in which each engine has its own fuel system on its side of the aircraft, composed of several interconnected tanks that function for all practical purposes as a single large tank.

Again, engine fuel control systems are simple, usually consisting of a single selector with only OFF, ON and CROSSFEED positions. In fact, in some systems, there are only OFF and ON positions, with a separate control to open or close a crossfeed valve.

Fuel Transfer Pumps

While these "many-into-one" multitank systems remain very simple in normal operation, they usually have certain constraints and failure modes not found in a real one-tank situation. This is due to the fact that often, the arrangement of the tanks in the airframe precludes simple gravity flow from one to the other at all times, so transfer pumps are needed to help out (Fig. 3-3).

In any of these systems, all fuel feed to the engine comes from one particular point in the multitank system. This is usually, but not always, at the low point of the system. Generally, some sort of hopper tank or fuel chamber is built in to make sure there's always an adequate head of fuel above the pickup point. Fuel from other tanks flows into the hopper tank, where it's picked up by suction, either from a separate boost pump or the engine-driven fuel pump. However, while the system has *boost* pumps to supply fuel to the engine(s), is also has *transfer* pumps to move fuel from one tank to another.

In some turbine aircraft, the fuel flow from the main boost pumps is sufficient so that some can be tapped off to provide "motive flow" which actuates so-called venturi or ejector pumps for low-pressure fuel transfer functions. You may have used similar pumps hooked to a garden hose to pump out a flooded basement or empty a swimming pool. A high-pressure flow directed through a small orifice creates a surrounding low-pressure area which entrains a large volume of liquid and "takes it along." In other installations, including most recips that use transfer pumps, electric

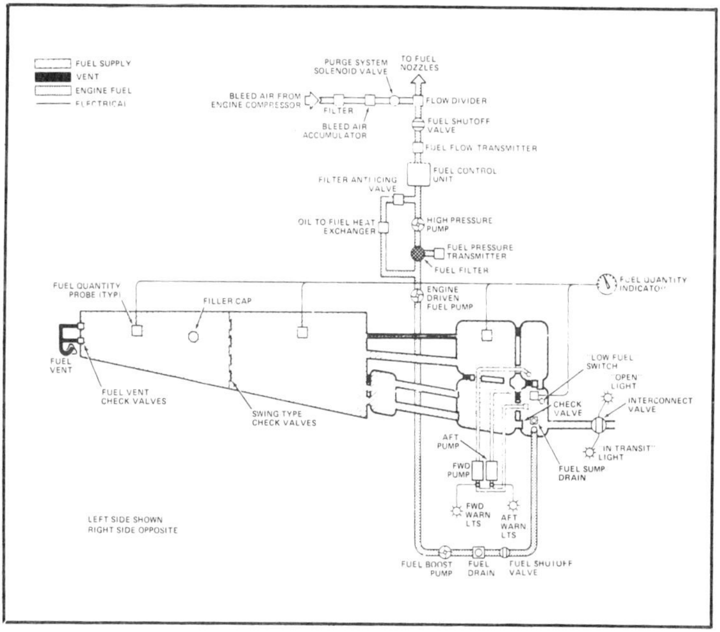

Fig. 3-3. In this Rockwell Commander system, fuel flows by gravity, assisted by transfer pumps, from the wing into the center section. From there, electrical and engine driven boost pumps move it into the combustion chambers. An identical system is on the other side of the aircraft, and is accessible for single-engine operation via an interconnect line and valve.

pumps pick up fuel from low points in the system and dump it into the hopper tank; any excess simply overflows back into the rest of the system.

The weakness of such systems is that they depend on the transfer pumps for at least a portion of their fuel. Thus, if a transfer pump fails, the pilot may not even notice it until he or she finds that some fuel in a given tank may be unusable, which can be a nasty surprise at the end of a flight when reserves may be scant. This becomes particularly nerve racking in systems where there's only a single fuel gauge for each side of the airplane, with no way of monitoring the contents of individual tanks.

While we're on the subject of pumps, let's consider the function of boost pumps. These electric units are located either in the fuel tanks themselves or in the lines between the tanks and the engines, and they can fulfill various important functions.

The most important, of course, is to provide fuel pressure in the event the mechanical, engine-driven pumps fail. It's important for the pilot to know the capabilities and limitations of the boost pumps in a given fuel system. For example, in some aircraft, the pumps in a given fuel system. For example, in some aircraft, the pumps operate at low speed continuously, switching automatically to high speed if an engine pump fails. In others, they must be switched by the pilot. Even then, in some recips the electric pumps cannot provide sufficient fuel flow for full-power engine operation. A pilot who blithely advances the throttles may find himself faced with detonation and burned valves, not to mention power loss, due to an excessively lean mixture.

Fuel crossfeed operations also can depend on boost pump functions. Some aircraft have suction crossfeed systems in which the mechanical and/or boost pump of the *operating* engine *draw* fuel *from* the inoperative side. Others have pressure crossfeeds, in which the boost pump on the *inoperative* side *feeds* fuel *to* the operating engine. It should be noted that in pressure-crossfeed aircraft, an engine can be kept running even if *both* its engine-driven pump and the boost pump on that side of the aircraft fail, although all fuel used will come from the operating-engine side. In some aircraft,positioning fuel selectors for crossfeed energizes and de-energizes the appropriate pumps automatically, while in others, this is the pilot's responsibility.

Generally, it's the newer aircraft that have the interconnected-tank systems (Fig. 3-4). As we examine the older ships, we find the more conventional multi-tank systems with a fuel selector position for each. In some of these, each engine has access to every tank on the airplane. In others (more common), each engine's fuel selector has positions for the tanks on that side of the airplane, plus a crossfeed position that offers access to the other side's fuel selector and whatever it's set to.

Obviously, the more tanks, the more possibilities for pilot error, and some of the older "plumber's delight" systems require considerable pilot management and careful consideration before moving any handles to avoid that embarassing feeling of " . . . it's quiet out there—*too* quiet . . ."

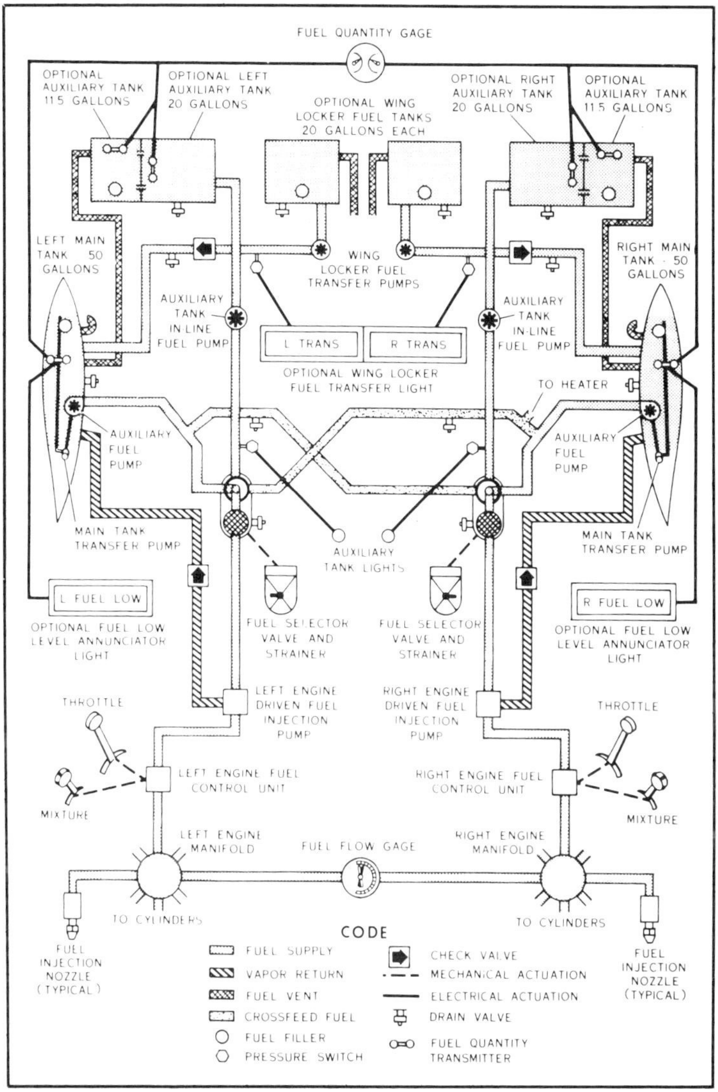

Fig. 3-4. The fuel system of this Cessna 340 (similar to that of earlier 421's) can incorporate as many as six tanks, two selectors, two transfer switches, and six electric pumps. Careful management is required to make all fuel available in any situation and prevent improper transfer from venting fuel overboard.

For example, consider the fuel system of an older-model Cessna 421. Each side's fuel system consists of a main tank (which, paradoxically, is out on the wingtip), an auxiliary tank in the wing, and an optional wing-locker tank in the engine nacelle. Each engine's fuel selector has positions for the main or aux tank on its side, and the main on the other side (as well as "off"). Electric fuel boost pumps are submerged in the main tanks; two more pumps are installed in the lines from the aux tanks to the fuel selectors. The nacelle tanks have no direct connection to the engines; instead, electric transfer pumps dump their fuel into the main tanks. Aux-tank fuel cannot be cross-fed.

Sound simple? Hold on, there are a couple of little quirks with this system.

Possible Fuel System Hangups

One is the problem of fuel and vapor return. The fuel injection system of a recip engine runs considerably more fuel through its high-pressure pump than the engine actually requires at any point except at the highest power. The excess fuel is returned to the fuel tanks. In the case of the Cessna under discussion, it's returned to the main tank on that particular side, regardless of fuel selector position. (In some other twins, fuel is returned to the tank from which it was drawn.) This means that the airplane must be flown for a certain amount of time to "draw down" the fuel in the mains before switching to aux fuel. Otherwise, the returned excess will have no place to go, and the main tank will overflow, dumping fuel overboard through the fuel vent lines. It's not just a few drops, either. Generally, a typical system draws almost twice as much from the tanks as is actually consumed. Another idosyncracy that results from this is the fact that the aux tanks appear to deplete at an alarming rate. If you're running a power setting that burns 12 gph, you'll still see those aux tanks dropping at 20 gph or better, and since the fuel gauges on the panel only indicate what's going on in the tank actually selected at the moment, you have to take it on faith that the remaining 8 gph are going into the mains (or, if you forget to burn them down first, overboard!)

What about a pump failure? Well, the small electric pumps in the aux tanks are designed for cruise operation only (incidentally, they have no separate control, but come on when the main tank boost pump switch is in "LO"). Should an engine-driven pump fail while running on the aux tank, it'll probably be necessary to switch back to the main tank on that side so that its larger electric pump will take over the job. Any fuel left in the aux tank on the failed side will be usable.

How about the nacelle tanks? You have no way of knowing how much they hold except by monitoring the fueling process. To use the fuel in them, you have to wait until you have sufficient room in the mains to accommodate both the contents of the nacelle tanks, *plus* the expected fuel return, and *less* the expected burn rate; then turn on the nacelle transfer pump switches. A light indicates that transfer has been completed (or that the nacelle tank is empty) and you monitor the main fuel gauges to double-check. If a nacelle transfer pump fails, that fuel is unavailable for either engine.

Admittedly, I've described one of the most complex fuel systems going, and Cessna has done a magnificant job improving the situation. The fuel system of the later 421's is a delight although they still have those darn nacelle tanks. But there are a lot of old 421's out there, and they're not uncommon as entry-level steeds for the aspiring corporate pilot. A typical fuel-use plan for a 421, then, might read as follows for a long-range flight: (Remember, what we're trying to do is keep the maximum number of options available for the maximum time, so let's try to use the fuel that might not be available to both engines as early as we can.)

- Takeoff and climb on the main tanks; the period of operation at high power and rich mixture will help make room for future fuel return.
- The aux tanks (with optional long-range system) are about the same size as the mains. Remember, though, that up to half of the contents will be returned to the *mains*, so let's wait until the mains are burned halfway down before we switch to the aux tanks. They'll appear to drop rapidly, but an occasional look at the contents of the mains (using the fuel gauge override switch) should show the main-tank level gradually *increasing* from the return flow.
- When the aux tanks are almost depleted, we'll switch back to the mains, which by now are almost full again! We're now in a desirable situation—any fuel aboard the airplane can be made available to either engine. Moreover, we're done with the feed-from-one-return-to-another situation; from now on, we'll feed only form the mains, and look only at the main-tank contents with the fuel gauge.
- When the mains have burned down a bit again, we can start transferring nacelle-tank fuel. Each nacelle tanks holds 26 gallons, so if we wait until the mains are half empty we're unlikely to get any overflow. (Actually, the nacelle tanks transfer at about 30 gph; assuming we're burning 15 gph from the mains while transferring; we'd be safe if we just waited until we had about 20 gals. "headroom" in the mains.)
- Last, but not least, most 421's have an air-conditioning pack installed in the right nacelle, and only in one nacelle tank—the left one. Thus, when we're done transferring nacelle fuel, we'll have 26 extra gallons on the left side. This will have to be crossfed by positioning the *right* fuel selector to the *left main* position and carefully monitoring both fuel gauges until they show the same amount.

Fun, huh? Now, imagine doing this sort of thing at night, on the dials, in weather, while coping with deteriorating weather at your primary destination, with the prospect of a hold at the alternate because everybody else is going there, too, and an amended routing from center. As I've said, the twins we fly today have the performance, sophistication, and some of the complexity of late-generation piston airliners. Are you starting to see why the liners carry a flight engineer?

Vacuum Air Systems

Let's take a break with a relatively simple system before plunging on into the wonderful world of electricity. Even the larger twins and turbo-

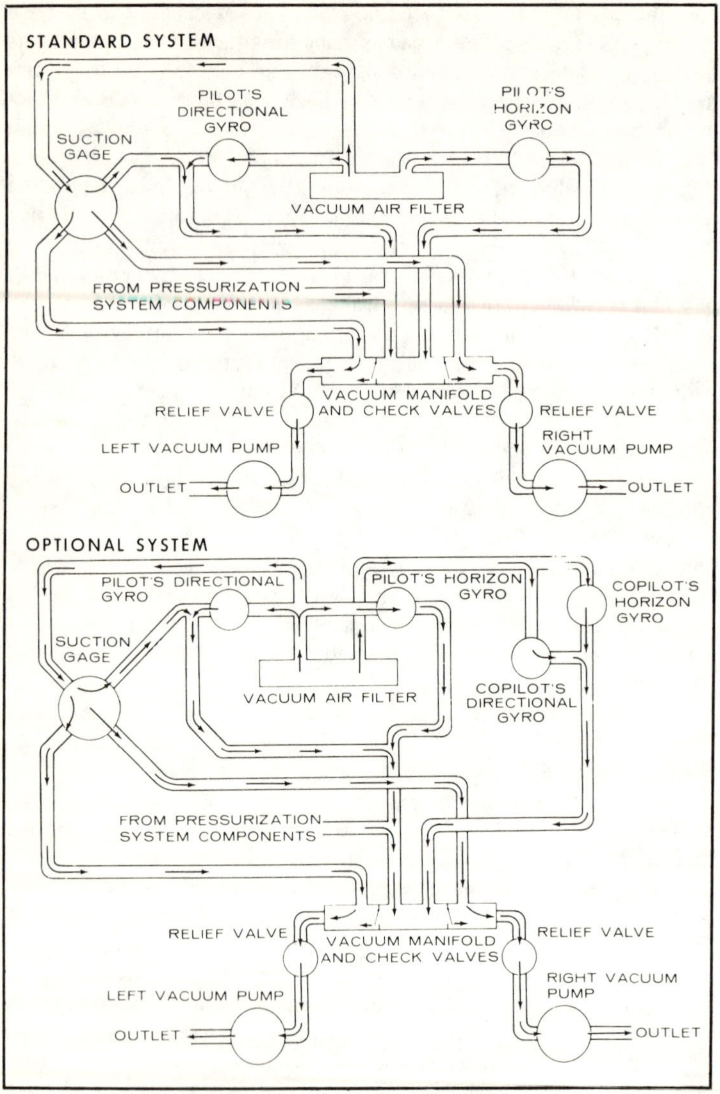

Fig. 3-5. Standard and optional instrument vacuum systems on a piston twin include a manifold with check valves for automatic fault isolation if a single source fails. Pressure systems work exactly the same way, although the check valves would be reversed.

props, which usually use electric gyros on the pilot's side of the panel, generally have air-driven ones on the copilot's side. Air is also used for de-ice systems (covered in detail elsewhere) and for inflating the door seals of pressurized aircraft. Luckily, most of these systems are pretty similar; about the only difference is that some are vacuum systems while others are pressure types, and the source of air pressure or vacuum may differ with aircraft types (Fig. 3-5).

The turbine aircraft rejoice in a virtually unlimited supply of hot air. Thus, each engine is normally provided with an air bleed, in which air is tapped off for a number of purposes including pressurization, air conditioning, heating, inlet de-icing, and, last but not least, instrument air (Fig. 3-6).

If we're dealing with a pressure system, all that's required is a pressure regulator and a heat exchanger to cool the air down a bit. If

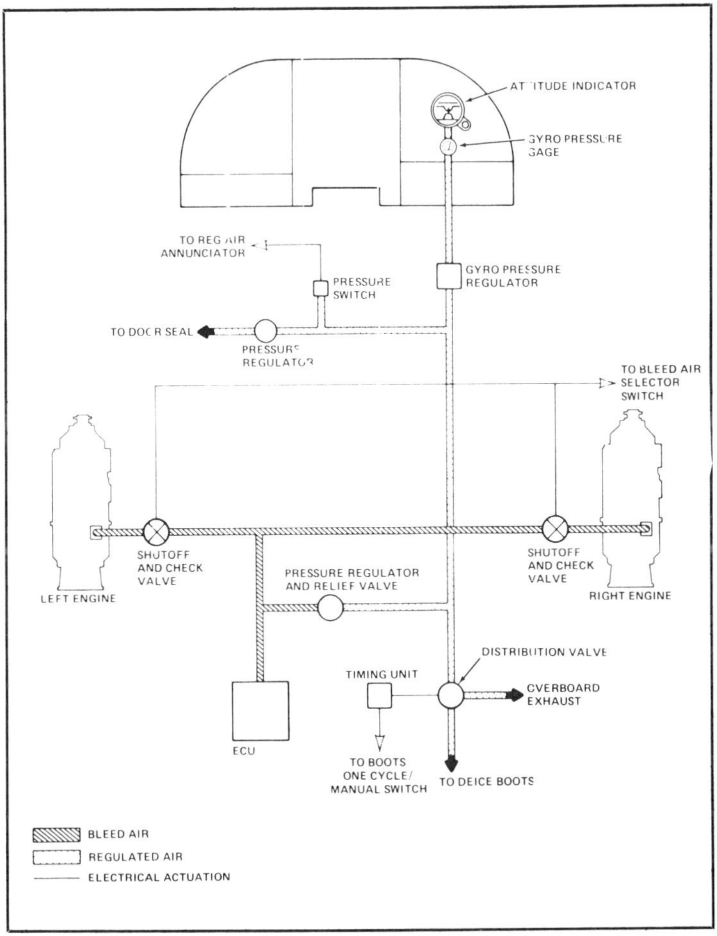

Fig. 3-6. In turbine-powered aircraft, instrument and de-ice air pressure are provided by the engine bleed system. If instrument vacuum is required, a venturi pump is used. Note that in this Rockwell Commander, the only pneumatic instrument is the copilot's attitude gyro; his directional gyro, and both the captain's gyros, are electrically powered.

vacuum is required, a simple venturi pump like the ones described in fuel systems, with no moving part, can do the job.

Recip-powered airplanes use either a "wet" oil-bath, or "dry" mechanical pumps geared to the engine. Either type of pump produces both vacuum and pressure; "wet" types, which are becoming less and less common, require oil separators to remove the oil from the outlet side of the pump. Generally, dry types are used in applications in which pressure instruments are installed, since no elaborate filtration is required to keep oil from the instruments.

In any case, the pressure or vacuum supply from both engines is ducted to a central manifold, usually behind the instrument panel, which contains check valves. These valves serve to isolate a failed source with no pilot action required, that is, if one air source fails, it will be shut off so that the other source does not have to pull (or push) air through the failed source or pump. The common air manifold is also the point at which the system pressure gauge is connected. Some older aircraft have a selector valve which switches this gauge from one source to another. Later types usually use separate indicators of some sort to indicate the relative health of each air source. A master vacuum or pressure regulator is connected either to the manifold (if air is used only for instruments) or to a separate instrument line if higher-pressure air will also be used for other purposes.

Some systems, incidentally, have no separate indicator for individual air sources. A simple way of checking them is during startup and shutdown. Start the engine first to determine whether its air source is operating properly (and that the one on the inoperative engine remains isolated). Then, after the flight, shut down that same engine first to verify the operation of the system on the other side. The same technique works for hydraulic systems.

While vacuum/air systems are simple ones, this discussion has brought up two important concepts: *redundancy*, or the ability to continue flight after the failure of a particular component, and *fault isolation*, the ability to keep a single failure from influencing other interconnected units. Both of these concepts are important in the area we'll consider next—electrical systems.

Electrical Systems

If there's any one system that we'd have to consider the lifeblood of an advanced twin, the electrical system would be the one. Not only does it fulfill its own required functions and, in this age of increased ATC and navigational complexity, those functions alone are the cornerstone of almost any flight operation, it also ties together virtually all the other systems in the airplane. Electricity operates the fuel systems, hydraulics, engine controls, de-ice equipment, environmental control, and many other devices on the airplane. Given the traffic and weather conditions in which we fly, a total loss of electrical power would be tantamount to disaster.

A discussion of the electrical system of any twin immediately leads to the subject of redundancy. In fact, it's as much the redundancy of systems as the redundancy of power plants that leads many buyers to consider a twin in the first place. When you're IFR in heavy weather, it's not just that second engine that brings peace of mind, it's that second generator, that second vacuum system, and that second hydraulic system. Electric power is so vital, that the manufacturers of some high-performance singles are currently hanging a second, smaller electrical system on that one engine. Lose all the knobs and dials when you're IFR under certain circumstances, and you're in deep, deep trouble no matter how many engines you have.

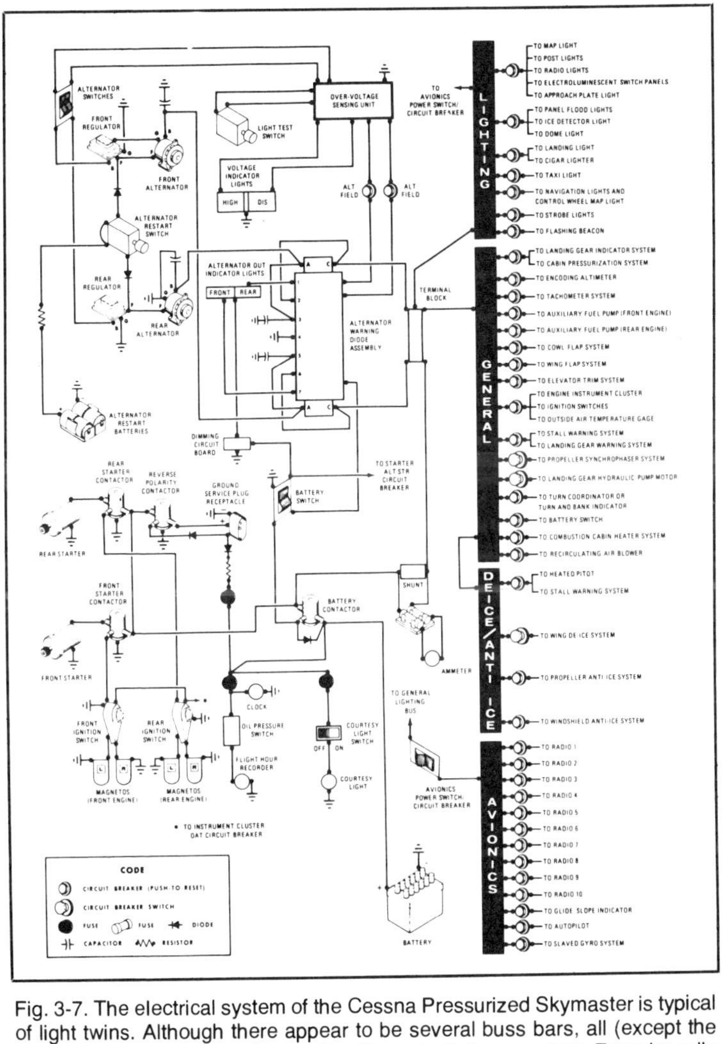

Fig. 3-7. The electrical system of the Cessna Pressurized Skymaster is typical of light twins. Although there appear to be several buss bars, all (except the avionics bus) are tied together, acting like a single buss system. Four dry cells are provided to restart the alternators in case of electrical failure.

The electrical systems of the medium and heavy twins vary from rather simple ones in some recips, similar to those of singles and light twins (Fig. 3-7), to highly sophisticated split-bus installations in some of the heavier recips and turboprops. As with other systems, greater complexity brings with it more redundancy and, especially with electrical systems, a high degree of fault isolation.

Let's start with the method of generating power. In most more or less modern recips, power is created by an *alternator.* This device is similar to a generator, but there are a couple of differences. One is that the alternator produces alternating, rather than direct, current; the alternating flow is *rectified*, or converted from AC to DC, before being fed to the aircraft's electrical system. Another, and more critical, difference is that, unlike a traditional generator, an alternator requires a DC voltage across its field windings to start producing current. In a properly-designed electrical system, such voltage is available from the battery, but it should be kept in mind that if the battery is completely discharged or otherwise unavailable for power and there's no other power on the system, an alternator, once shut off, *cannot be restarted* without some means of temporarily energizing, or "flashing," the field.

Turboprops still use generators rather than starters. While a generator is inherently less efficient and heavier than an alternator, it has one important plus for turbine applications—feed power *into* it, instead of drawing power from it, and it becomes a DC motor suitable for use as a turbine starter, and while it's heavier than an alternator, it's lighter by far than a separate alternator and starter. Thus, turboprops use a *starter/generator* unit, and their electrical systems reflect the changes needed to make these units perform in their double role.

Now that we have our two redundant power sources, we have to have some means of controlling them and distributing the power they produce. Here's an area that offers further possibilities for redundancy and fault isolation.

Generators and alternators must be controlled to produce the correct voltage, using a device called a "voltage regulator." Obviously, each generator or alternator needs one of these, but there are a couple of ways of hooking them up. Older, lighter twins in particular, which use generators rather than alternators, are often found with an arrangement in which one voltage regulator runs both generators, with the second one available to be switched in as a reserve. Alternators, on the other hand, have to have their own controllers. Usually, there's a *paralleling* provision connected between the two voltage regulators so that the alternators share the load more or less equally. Such paralleling is best checked during runup. Turn on enough electrical equipment to approach the limit of a single alternator or generator's capability (e.g. landing lights, air conditioning, etc.) and note whether both units show about the same output in amps. Now, turn off one generator or alternator and note that the remaining one takes up the entire load; then reverse the sequence.

Together with control of generator/alternator output, we must examine the possibilities for protection should a unit go out of control. Both over- and undervoltage can be harmful—overvoltage can roast avionics and start thermal runaways and fires in Nickel-Cadium batteries, while undervoltage can cause battery discharge, generator damage from reverse current, and popping circuit breakers, not to mention inadequate performance from such vital aids as fuel boost pumps and de-ice equipment. Most installations incorporate an overvoltage protector which will drop a generator from the line if the voltage rises above the safety limits. Most installations also have sensors that indicate insufficient output.

It's in the area of power distribution that larger aircraft make best use of their potential for fault isolation. Smaller aircraft, including many of the pressurized recips, use a single bus system that's similar to those found in singles, or, for that matter, in your family car.

In these systems, all the electrical sources, that is, the two generators or alternators and the battery, are connected to a single main conductor, or busbar, as are all the electrical loads. Individual circuits are protected to some extent by fuses or circuit breakers (Fig. 3-8), which open the circuit if there's an excessive current drain.

The weakness of such a system is, of course, that the single bus itself represents a single point of failure. As long as it's only individual devices

Fig. 3-8. Circuit breaker panel of a Super King Air (additional breakers are located on the instrument subpanel). Note how the panel is divided by functional groups, with engine and electrical breakers further subdivided into left and right groups. All avionics breakers are grouped at the right.

that fail, there's not too much of a problem, their circuit breakers will protect the integrity of the system as a whole. The problem is that any failure of the bus itself, and that could be something as insignificant as a screw working loose behind the panel and dropping between a hot bus connection and a ground, will result in what is, for all practical purposes, a total electrical failure. If the bus shorts to ground, neither the battery nor either generator can apply power to it. By the same token, if the connection between a generator and the bus fails, that generator is useless, there's no alternate route for electricity.

The Rockwell Commander 840 System

As a contrast, let's examine the electrical system of a typical turboprop—in this case, the Rockwell Commander 840 (Fig. 3-9). Power from the dual lead-acid batteries and the dual 300-amp output starter/generators is applied to a single main bus. This is the only single bus in the airplane. (On some other airplanes in this class, the main bus is further divided into a battery bus and a left and right generator bus interconnected by current limiters.)

From the main bus, power flows through circuit breakers to five other buses: left and right distribution buses, left and right radio buses, and a cabin bus. This last bus carries only nonessential cabin equipment (e.g. galley stove, stereo, etc.) and is protected by fuses.

All essential electrical users are connected to one or the other of the distribution buses. Avionics are more or less equally divided between the radio buses. Each distribution bus offers two alternate power sources: a "reset" circuit breaker which connects it to the main bus, which is normally closed, and a *bus tie* circuit breaker, normally open, which connects to the opposite distribution bus. Moreover, each distribution bus is further subdivided into the portions which feed the overhead panel and those which feed the lower circuit breaker panel; a "power" breaker is between these two sections.

Thus, for example, a failure in the overhead section of a given distribution bus can be isolated to that section alone. Let's say that the left (#1) overhead distribution bus shorts out. The "reset" breaker will drop out, the "bus off" light on the lower circuit breaker panel will illuminate, and everything powered by the #1 bus will shut down.

All is not lost, however. First of all, the airplane's essential services are more or less equally divided between the two buses, so even if #1 distribution bus were to remain down for the rest of the flight, it would be merely an annoyance rather than an emergency.

In your case, the pilot doesn't know yet where the fault lies, so he or she manually trips the *power* circuit breaker to isolate the overhead portion of #1 bus from the circuit-breaker panel section. Next, an attempt is made to reset the "reset" breaker. If it refuses to reset we now know that the fault lies on the overhead (and/or auxiliary) section of #1 bus. Leaving the "power" circuit breaker tripped isolates the lower section of #1bus from

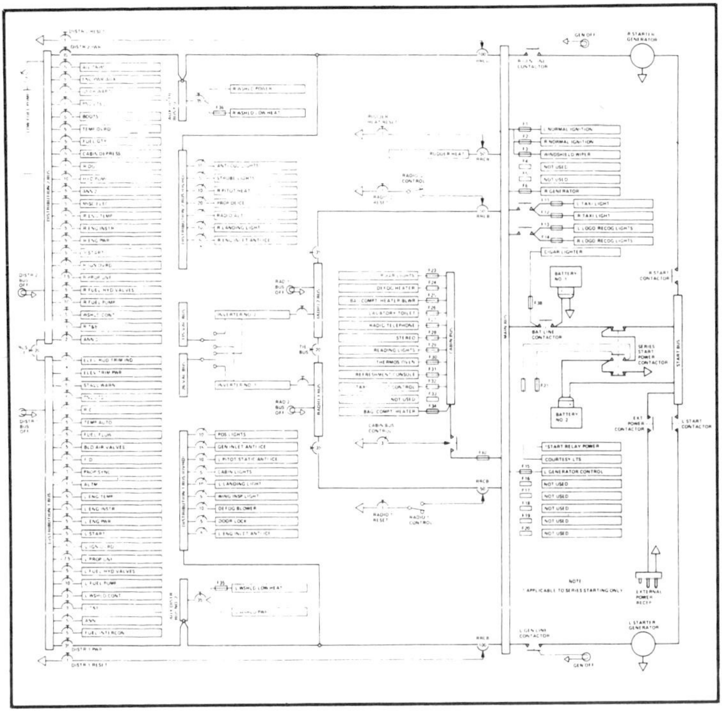

Fig. 3-9. In the Rockwell Commander 840, busses are interconnected by circuit breakers, allowing manual fault isolation and selection of alternate power paths. Since this aircraft is powered by Garrett engines, switching is provided to allow 48-volt starts on cold days (series start power contactor).

the shorted overhead section. We can now close the "bus tie" circuit breaker and re-power the lower section of #1 bus via its alternate path from #2 distribution bus. Thus, rather than losing all electrical services due to a bus failure, we've lost only the following: nav lights, left engine inlet anti-ice, cabin reading lights, wing inspection lights, the pilot's heated pitot (but the copilot's still works,) and a couple of relatively minor services like defog blowers. Everything else in the electrical system continues to work normally, as do all the avionics. From a practical point of view, the airplane has four separate distribution busses, and there are alternate current paths available such that any one of them can be isolated while the other three remain powered.

A similar split-bus system provides power for the avionics. Each of the two radio buses can be powered from the main bus, *or* from the opposite radio bus. Moreover, each radio bus carries an identical inverter for aircraft AC power, and the 115 VAC and 26 VAC busses that power such systems as radar and parts of the autopilot can be switched to either inverter.

The whole idea, then, behind complex electrical systems is to arrange them such that single points of failure are minimized or eliminated. This "suspenders and belt" philosophy allows continued operation, admittedly with some degradation of capability in certain failure modes, in situations that would cause immediate, complete, and unrecoverable electrical failure in simpler systems.

A few larger aircraft go even further with a so-called "essential power bus." This bus offers a means of alternate power for those few things you just plain can't do without, such as the most important boost pumps, the instrument lights, an electric gyro horizon, one nav radio, one com radio, and the transponder. The essential bus can be powered directly from the battery, with no other main or distribution busses involved. Some installations even include a separate small battery (much too small to handle engine starting loads), which is constantly being trickle charged by the airplane's electrical system when it's operating normally, and which can power the essential bus when all else fails.

There's another type of emergency battery supply that you may find on some of the recip twins that use alternators for power. You'll recall that I mentioned that an alternator needs a certain amount of field voltage to function properly. It's conceivable (although unlikely) that you might be operating with the battery totally discharged or otherwise offline, perhaps due to a failed battery contact, and/or a momentary extremely heavy electrical load, such as lowering the landing gear. This could drop the bus voltage to the point where either or both alternators would drop offline as well, resulting in total power loss. To remedy such situations, some alternator-equipped twins have some provision for providing battery voltage directly to the alternator fields, bypassing the usual battery contactor. Such devices are usually labelled *emergency alternator field, alternator flash,* or *alternator restart,* and their proper use should be understood by pilots of aircraft equipped with them. Of course, if the aircraft battery is completely and totally dead, it won't even provide enough of a kick to restart the alternators. A few twins offer even further redundancy by providing an entirely separate battery supply for the restart circuit. It doesn't have to be much, a couple of flashlight cells will do the trick. A friend of mine had both alternators go down on a Piper Aerostar after a battery failure (luckily during the daytime, and in VFR conditions on top) and was able to get things going again by feeding power from the batteries in his flashlight back in through the cigar lighter socket!

Hydraulic Systems

For many pilots, the transition into larger aircraft will mark your first introduction to the (usually) wonderful world of hydraulic systems. In lighter aircraft, the relatively smaller size of landing gear and flap system components makes it feasible for them to be operated by simpler means, such as electric motors and mechanical linkages. As we deal with larger and larger aircraft, though, hydraulics become a more attractive method of providing "muscle." In fact, by now a number of new smaller aircraft are using hydraulics for landing gear actuation (although they usually use electric, rather than engine-driven, hydraulic pumps). This is because it's easier to route hydraulic lines when you're laying out the design for a new ship. These lines can go through little holes and around corners, as opposed to moving linkages which require clearance to operate, bellcranks to go around corners, and all that sort of stuff.

In essence, a hydraulic system isn't that different in concept from an electrical system. In both, we require a prime mover, or source of power. In the electrical system, the source of power consists of generators or alternators, in the hydraulic system, it's pumps. We need a means of controlling the power—switches for electricity, valves for hydraulics, as well as wires or tubing, to deliver it to where it'll be used. One difference is that every hydraulic "user," such as a gear or flaps, requires not a single wire of an electrical connection, but *two* hydraulic lines, since fluid must be returned to the low-pressure end of the system as it's used.

There are two basic types of hydraulic systems: constant-pressure, in which the entire system is pressurized whenever an engine and its associated hydraulic pump are running, and the more modern *demand* types in which only a small part of the system, the pumps themselves and an accumulator, are under pressure except when actual hydraulic loads are applied. Such systems are more common nowadays, and it's a good thing, too. The constant-pressure types tend to be real maintenance hogs, since the airplane is filled with yards and yards of tubing through which hydro fluid is circulating merrily at pressures of up to 3000 psi, just looking for a way to get out! Demand systems, too, tend to operate at lower pressures, usually around 1000 psi, which not only saves maintenance, but allows the use of lighter components. Let's examine a typical system.

Examining a Hydraulic System

We'll start with the prime movers—the hydraulic pumps connected to the aircraft engines. Nowadays, it's more or less standard practice to put a pump on every engine you have, but this hasn't always been the case. Those of you who learned your multiengine flying in an Aztec or Apache, or certain older Commanders, will no doubt remember those anxious moments after losing the engine near the ground with the single hydraulic pump. I'm sure you'll never forget those busy moments during which you flew the airplane with one hand, secured the dead engine with the other hand and flailed wildly with the hand hydraulic pump with whichever hand you could free for a moment.

Both pumps draw their supply of hydraulic fluid from a common reservoir. This reservoir, of course, should be a stop on your preflight inspection, and if the system is ever serviced, make sure that it's refilled with the correct sort of hydraulic fluid. Most aircraft use Mil-Spec. 5606 red fluid, but check it to make sure. Fluids and seals of different types are incompatible, and the wrong kind will immediately ruin all the rubber components in your system!

In those systems, in which more than just the landing gear is hydraulic, there's almost invariably some sort of backup pressure source to deal with the unlikely eventuality of both engine-driven pumps being unavailable. This, too, draws its fluid supply from the same reservoir, but with an added feature—the backup supply comes from a separate standpipe in the reservoir which will retain a quart or so of fluid even if the main reservoir is depleted. Why? Because a break in the primary pressure line from either pump to the rest of the system could speedily pump all the fluid into the reservoir overboard, leaving you with no fluid for the remaining pump to use. (Actually, the remaining pump might well fail shortly thereafter anyway, since most hydraulic pumps rely on the fluid itself for lubrication, and won't run dry for too long. This, by the way, is why really large airlines have several completely separate hydraulic systems, each with its own reservoir and pumps.) In some aircraft, the emergency pump is the traditional big red handle. In others, it's electric. In the system we're examining here (Rockwell Commander 840), the emergency system can supply the flaps, power brakes, nose wheel steering, and parking brake (Fig. 3-10). Note that there's no provision to power the gear by the emergency system, either to retract it or extend it. Other means are provided for emergency extension, because the apparent philosophy of the designers was that dual hydraulic failure immediately after takeoff would be quite unlikely. A final word on emergency systems—some (especially older Commanders) include a priority valve by which the pilot can isolate certain system components after a failure. For example, if you lost hydraulic pressure in a 560E and pumped the gear and flaps down by hand, you could then select the "brakes only" position on the priority valve before landing, pump up as much pressure as possible in the accumulator, and know that you'd have sufficient pressure for several good brake applications without wasting any on such frills as nosewheel steering.

All right, let's get back to the main system. While the common reservoir supplies both pumps, it does so via two separate supply lines, each with its own shutoff valve. This enables the pilot to interrupt flow to either pump in the event of an engine or nacelle fire. Unfortunately, in most airplanes nowadays, this cutoff function is wired together with the firewall fuel cutoff, which eliminates one option for fault isolation. If you have a separate hydro cutoff and can determine that you have a leak between the pump on one side of the aircraft and the accumulator, you can shut off that pump only and save most of the fluid, as well as having hydraulic pressure available in the rest of the system.

From the pumps, the high-pressure fluid flows to a common hydraulic accumulator, a device analogous to the battery in an electrical system. Think of it as a pressure tank, or a "wide spot" in the system in which a certain volume of fluid is stored at the system's working pressure. In a

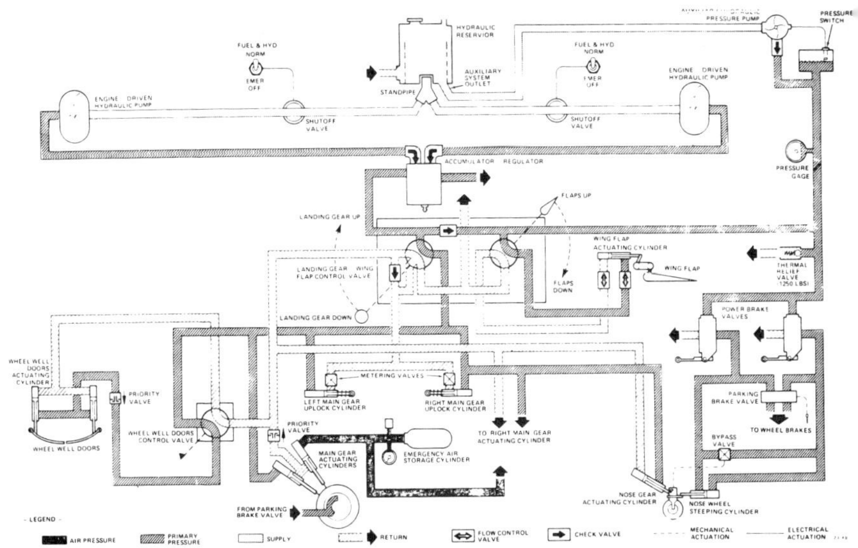

Fig. 3-10. The hydraulic system of the Rockwell Commander 840 incorporates two engine-driven pumps and a backup hand pump. An emergency air storage cylinder aids free-fall emergency landing gear extension and locking. Note how the power brake valves also control nose wheel steering.

demand system, the pumps are connected to the accumulator through a pressure regulator, or *unloader valve,* which not only controls system pressure between certain limits, but allows the pumps and pressure tubing to "relax" when not needed. Let's say our nominal system pressure is 1000 psi; when the pressure in the accumulator rises to about 1100 psi, as it would if the pumps were running and no load was present, the valve would open and dump the fluid flow right back into the reservoir! By the same token, if pressure drops to around 900 psi, the valve closes and allows the pumps to pressurize the accumulator once again. Thus, we can consider the hydraulic system to be composed of two interconnected "loops"—the *primary loop,* from reservoir to pumps to accumulator and back to the reservoir, and the *secondary loop,* from the accumulator to the various users, such as flaps, gear, etc., and back to the reservoir once again. The system pressure gauge is connected to the accumulator, and should always show the correct working pressure, give or take a hundred pounds or so. Pressure in the primary loop can vary from full accumulator pressure (when the system as a whole is in use) to almost nothing (when the unloader valve is open, the pumps are "loafing," and fluid circulating in the primary loop has almost no resistance.)

Landing Gear Systems

There are only so many ways to skin a cat, and there are only so many ways to make wheels go up and down with hydraulic cylinders. There are a few subtle differences between different types of systems, though, and

Fig. 3-11. The forward-retracting main gear of the Super King Air is held down-and-locked by the over-center position of the drag link (center of picture). Note the small links on the side of the main gear forging, which actuate the doors. Dual wheels allow the use of smaller diameter tires for full enclosure of the retracted gear inside the nacelle and provide location for dual brakes.

knowing how yours works, both normally and in its various failure modes, may someday help you avoid expensive scraping noises.

All systems are similar in certain points: a hydraulic cylinder is connected mechanically to the gear leg at some suitable place; ducting pressure to one side of it, while allowing fluid trapped on the other side to return to the reservoir, extending the gear (Fig. 3-11). Reversing the connections will retract the gear. Some form of downlock is invariably provided, and this is independent of the hydraulic system once it is locked. The downlock is usually some sort of over-center device. Where the differences show up is in how the gear is locked when it's retracted (if at all), how the gear doors are activated, and the types of provisions that are provided for emergency gear extension.

In a number of systems, there's no mechanical uplock as such. In some, hydraulic fluid is trapped in the *gear up* line by a valve when the gear is retracted; while in a few older constant-pressure systems, system pressure actually shoves the gear up all the time. In these latter types, of course, a loss of hydraulic pressure will result in the gear coming out of the uplocks, and possibly extending. In demand systems, it'll normally stay retracted, but leaks or deteriorating o-ring seals can allow it to "bleed down." If your airplane has a strange tendency to develop a "heavyish" wing on longer flights, or sometimes needs unexplained rudder trim inputs, you might want to check what type of gear it has, and if necessary, have it jacked up and checked for the *bleed-down syndrome.*

Most other systems incorporate a mechanical or hydraulic uplock. Generally, hydraulic uplock systems are designed so that system pressure holds them in the "locked" position, but I know at least one airplane that requires hydraulic pressure to *pull* the uplocks. In that airplane, a complete loss of hydraulic *fluid* would leave the pilot with no way of extending the gear.

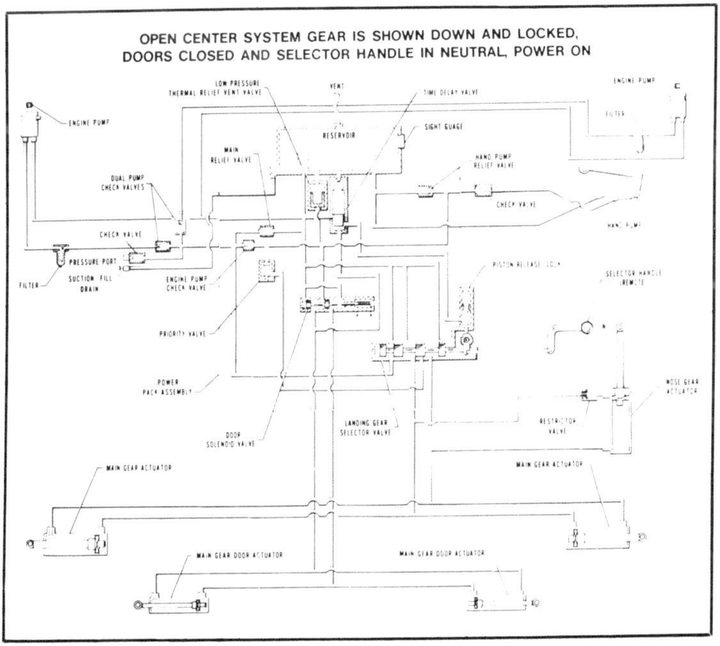

Fig. 3-12. The piston release lock unit in the hydraulic power pack of this Piper Navajo system causes the gear handle to "spring back" to a neutral position after gear cycling is complete.

Many aircraft also incorporated hydraulically operated landing gear doors. These require some way of sequencing them to make sure they're open whenever a wheel's about to zip past. Some use mechanical valves operated by the gear itself as it moves in and out of the well, while others use solenoid valves operated by microswitches. Some aircraft even use pressure sensing devices to monitor pressure as it builds up in the gear actuators. It's a good idea to know how the one in your airplane works. In some cases, a hanging gear door can be a prelude to more serious hydraulic problems. By the same token, many a flight has been aborted, and expensive mechanic types summoned, because a pilot saw a hanging gear door and didn't realize that the cause could have been found in an apparently unrelated spot, such as, the circuit breaker panel. It goes without saying, of course, that continued high-speed flight with a hanging door can give new and real meaning to the old phrase, "blow your doors off,"—and should you crumple a door, the bent remains might lodge somewhere that could be very embarassing at gear-extension time!

This line of thought leads us naturally toward the idea of emergency extension systems, and, again, knowing your system can put you in a good position when the chips are down (and the wheels aren't). This is particularly important when we consider that some emergency systems are of the "last resort" variety, and significantly reduce your available options if and when you're forced to use them.

The simplest emergency systems, and those easiest to use, are the ones found in airplanes in which hydraulic pressure holds the gear up (Fig. 3-12). As I mentioned earlier, in some of them the gear simply extends by itself anyway; on others, in which trapped hydraulic fluid forms the uplock, a simple "dump" valve relieves the pressure and allows the gear to free-fall into a down and locked position. It should be noted that airframe manufacturers have established specific airspeeds at which "free fall" extension is most efficient. Obviously, aircraft in which the gear retracts forward will be aided by airflow during an emergency extension, while those that retract aft will be hindered by it.

The next level of complexity is embodied by those aircraft that have uplocks. Here, some mechanical means (cables, etc.) are provided to pull the uplocks at the same time a valve dumps system pressure, again allowing free-fall. Note that in any of these systems that include hydraulically operated gear doors, the doors will generally remain open after an emergency extension, possibly requiring lower airspeeds (Fig. 3-13).

At least one aircraft I know of uses an *assisted free-fall* system. This is the Turbo Commander (in its various versions), which has large, aft-folding main gear. Airloads tends to help the mains retract and there's a nitrogen bottle plumbed into the system as a sort of "air spring." It contains nitrogen gas under moderate pressure, and there's a specific reading to check on preflight. As the gear retracts, it pressurizes the nitrogen to a higher value. If an emergency extension is required, placing the gear handle in the "down" position bleeds trapped fluid from the

Fig. 3-13. Double-wheel landing gear improves short-field performance; on a nose gear it also reduces steering effort. The "pip" pins on the torque "scissor" link enable it to be disconnected to allow 90-degree steering when the aircraft is being towed. The flush landing light extends when in use.

uplocks, which are spring-loaded to the "open" position. The nitrogen then forces the gear to extend and lock, although somewhat more slowly than normal.

Finally, a number of aircraft have so-called *blowdown* bottles. These are high-pressure cylinders of compressed air or nitrogen, which can be valved into the *extend* side of the gear system hydraulic plumbing. If all hydraulic power is lost, and the gear will not free-fall, placing the selector valve in the "down" position and actuating the blowdown bottle should extend it, and rather rapidly, at that!

The *warning* associated with the use of a blowdown is that the gas charge is, literally, your last shot. As the blowdown bottle pressurizes its lines, its gas moves shuttle valves in the gear system from the "hydraulic" to the "gas" position. Thus, the gear cannot be retracted until the aircraft has been serviced on the ground.

From a practical standpoint, this means that a blowdown bottle should be used with caution, and there are situations in which you might not want to use it at all! I might interject at this point that the blowdown will only work if it's got enough pressure to start with, which makes it an important preflight check item. Insufficient pressure will give it just enough pizazz to set the shuttle valves, but not enough to do you much good, landing-gear-wise.

The situation in which I would elect not to use the blowdown (in fact, one in which I didn't use it, a few years ago) is one in which one or the other main gear fails to extend. If this happens, and there's no indication of major hydraulic problems, it's likely that the gear is jammed mechanically. (For that matter, any hydraulic problem that keeps one main retracted is likely to affect the gas system as well.) Depending on what you're flying, you might elect to go in with all three gears retracted, rather than one main up and one down. If you try the blowdown, and the failed main is *still* retracted, you've run out of options—you can't retract the other gears anymore, so you'd better hope for a nice wide runway.

Personal Venture

The time I elected to avoid the blowdown was when I had a left main gear failure in an old Apache. As things worked out, I had 8500 feet of runway available, so after making a couple of passes to bounce the good gear on the runway in an attempt to jolt the left main loose, I climbed to 5000 feet above the middle of the field, retracted the gear, feathered both engines, nudged their two-blade props to a horizontal position with the starters, and spiraled down to a landing. Since the designers of the Apache considerately left the bottom half of all three wheels protruding when retracted, the only damage was the grinding down of the boarding step. I even managed to steer with the brakes. We jacked the airplane up (it's not easy getting jacks under an airplane that only sits about 7 inches off the ground!) and found a broken bolt in the left gear overcenter link. There was no way the blowdown bottle could have extended it.

Flap Systems

Not all that many aircraft in the classes we're considering have emergency flap actuation systems; it's generally feasible to find a runway long enough for a no-flap landing. Some recent airplanes, though, have enough gimmicks in their flap systems, whether hydraulic or mechanical, to warrant a few words (Fig. 3-14).

The simplest hydraulic flap systems are those with a single actuator and a simple *up-hold-down* control valve. The center position of the valve closes off both the up and down lines, so trapped fluid will hold the flap in whatever position it had when the valve was moved to that position.

A few airplanes have combination electric-hydraulic flap systems, in which the hydraulic solenoid valves are controlled by an electric switch on the instrument panel. The advantage of such systems is that they lend themselves easily to the incorporation of a pre-select feature. The pilot simply moves the handle to a desired point anywhere in the range, and when the flaps reach that point they halt automatically.

The interesting new wrinkle mandated by recent FAR changes is some means of preventing and/or dealing with the problem of asymmetric flap extension. Obviously, such a problem could cause very large rolling

Fig. 3-14. Use of spoilers (extended in this photo) instead of ailerons for roll control on the MU-2 allows the entire trailing edge of the wing to be used for double-slotted Fowler flaps for short-field takeoff and landing performance.

tendencies, possibly beyond the ability of the ailerons to counteract it. Various schemes are used, including limit switches or mechanical comparators. The asymmetry detection system will stop all flap movement if the difference between the two flap extensions becomes significant.

Pilots should be aware of this possibility. If you're in the middle of extending the flaps and they quit moving, don't just forge ahead and take steps to get them going again. They may have quit moving for a reason, and further attempts to either raise or lower them may make the situation worse. If it's possible to continue the flight to an airport and land with no further flap use, and if the rolling tendency isn't too severe, this is your best course of action. If a flap has jammed and the other flap can still be moved, you can carefully attempt to line them up visually (or by roll feel). Once you've done this, leave them alone!

Power Steering and Power Brakes

No, we're not talking about cars here! A few of the turboprops you might be getting into have these niceties, and even one light recip pressurized type (the Aerostar) has them, so let's devote just a moment to this last portion of our hydraulic considerations.

One difference between regular and power brakes is that the preflight is a bit simpler for the power-brake ships. Instead of a separate brake fluid reservoir, they use the same Mil-H-5606, or whatever that the main system uses. Of course, it takes a little practice to develop a light toe to use with them, just take it easy at first.

The operational difference is that, unlike regular brakes, power brakes can fail in *two* ways instead of one. Regular brakes can fail for various reasons, all related to their hydraulic and/or mechanical connections. Power brakes can enjoy all the same failure modes, as well as being subject to failures of the airplane system as a whole. No big thing, and, as we mentioned before, most emergency hydraulic systems retain ways of powering at least the brakes when all else fails, but you, the pilot, have to be aware of the situation. I've known at least one instance in which a pilot (admittedly of a larger aircraft than we're dealing with here) did a terrific job of dealing with a hydraulic failure inflight. He made a perfect no-flap landing after a perfect manual gear extension, and then he rolled off the far end of the runway because he'd forgotten to switch a priority valve for emergency brake power, and the pedals just went to the floor. It behooves you, then, to be aware of how to operate your emergency braking system, as well as its limitations. The next time you come back from a flight, after you shut down the engines (and de-energize any electric hydro pumps you might have), cycle the brakes a few times while watching your hydraulic pressure gauge, and see just how many brake applications you'd have in a real emergency situation.

Power nosewheel steering can make moving a heavy aircraft around on a crowded ramp a real delight. It comes in two basic styles—linked with the power brake pedals, as in Commanders, or operated by a separate control, as in everything else. Anyone making a transition into a Commander will spend a few minutes weaving along a taxiway in the so-called "Bethany Boogie" before learning the correct "Twinkletoes" touch necessary for a smooth ride, but it'll settle down. Just remember that nosewheel steering will not be available during a rollout after a hydraulic emergency in some cases, so be prepared with rudder, brakes and even asymmetric power and/or reverse, as is necessary.

In the case of the Commander (and its descendant, the Aerostar, which steers with a toggle switch on the throttle quadrant), it can be noted that the authority of a nosewheel steering system trying to turn a single large wheel on a dry ramp is limited. If the airplane is stationary, the wheel won't turn very far until you get underway. This is really an operational note, and it doesn't belong in a systems chapter, but if you're driving one of the late Ted Smith's remarkable twins, try to avoid coming to a stop with the nosewheel cocked all the way over, or it might take rude blasts of power to get things straightened out.

Larger twins, and the Merlin turboprops, use an airline-style nosewheel tiller mounted near the captain's left knee. Operation is instinctive, the further you turn the handle, the sharper it steers. Since double-wheel nose gears offer virtually no resistance to steering inputs, even when at a standstill, it's easy to make overly sharp taxi turns until you're used to the system's sensitivity.

De-Ice and Anti-Ice Systems

The class of aircraft which we're considering in this book includes many that can be equipped for flight in icing conditions, and many of them will be flown in ice fairly frequently. In later chapters we'll deal specifically with the operational aspects of icing flight and the inflight use of de-ice gear. Here, we'll just examine the equipment and how it works.

The most basic anti-ice device, of course, is one that you'll find on just about anything from a Cessna 150 on up: a heated pitot tube. More advanced ships usually have provisions for heating the static ports, stall warning vane, and possibly the fuel vents as well. A frozen fuel vent could shut off fuel from a given tank, or it could even cause structural damage as powerful boost pumps try to "suck a wing flat."

Even better than a heated pitot, is a *pair* of heated pitots! One of the nice things about the larger instrument panels in cabin-class airplanes is that there's room for copilot instruments, whether there's a copilot looking at them or not. When dual instruments are installed, there are two heated pitots, two static inlet systems, and so on down the line, and if there are split-bus electrical systems, normal installation practice is to power one

Fig. 3-15. Propellers are de-iced electrically in modern installations; power is fed to the blade boots by slip rings at the hub. A timer controls alternate application of power to inboard and outboard blade sections; the pilot should monitor the de-ice indicator to be sure that all blades are de-icing evenly, preventing disastrous imbalance. In this Piper Cheyenne installation, a heated boot also protects the engine inlet.

pitot system from each bus. Are you starting to understand the idea of redundancy?

Having started with pitots, which are generally at the front of the airplane, we'll work our way aft. On most twins, the props come next. Nowadays, electric heat is the universally accepted method of propeller anti-ice.

Note that word: *anti-ice,* as opposed to *de-ice.* The terms mean different things—anti-ice equipment *prevents ice from forming,* while de-ice equipment *removes ice that's already formed.* Some systems could do either: a good hot pitot heater, for example, can deal with ice already present, although it would normally be turned on well in advance to prevent the chance of losing the airspeed indicator (Fig. 3-15).

Props—Anti-Ice

Let's get back to our props, though. "Hot" props should be considered a strictly *anti*-ice measure, even though their heating elements could conceivably melt a certain amount of existing ice. Why? Because once ice has been allowed to form on a propeller, shedding it from one blade and not the others can result in catastrophic vibration. This is why prop anti-ice systems invariably include a little ammeter to monitor the current flow to each element. To make the best use of available electric power, each blade's heated boot is generally divided into two sections—inboard and outboard; a timer switches the current between them every 15 to 20 seconds or so. (It also switches between left and right props in turn on many aircraft.) Each time it switches, there'll be a momentary "kick" in the meter indication, but it should settle down within the narrow green band each time. If it ever reads high or low for one timer period, you'll know that some inboard or outboard element has failed, and you can expect the possibility of heavy vibration if that blade segment happens to ice up.

Earlier planes, and you may still encounter one or two during your career, used Demon Alcohol for prop anti-ice functions. Unfortunately, this was often used for *de*-ice functions as well. With only a limited quantity of alcohol available, pilots were often reluctant to start expending it until they were sure it was necessary, sureness coming from the shakes and thuds as ice started building up on the props and getting flung off against the fuselage. In our happy modern times, we seldom have to worry about running out of electricity. I generally heat up the props anytime I see 4° C or less OAT and expect to encounter any sort of airborne moisture.

Windshields—Anti-Ice

Moving further aft yet, we encounter a windshield. Here, too, older airplanes use alcohol, often in conjunction with the windshield wiper. This combination can handle light ice without too much trouble; just bear in mind, if you find yourself flying something with a whiskey windshield, that you'd better stop the booze flow a minute or so before landing (especially at night), or your vision may be very blurred.

Most modern aircraft use either electricity or hot turbine bleed air for windshield anti-ice. In some installations, bleed air systems have enough moxie to qualify for de-ice use as well. Like props, though, why should we force the issue? You're not going to run out of air or electricity. The time to turn on the de-ice equipment is *before* you need it.

Electric windshields come in several styles. The simplest and cheapest is an external "hotplate," a sheet of heated glass mounted outside the windshield and resting against its front surface (with a small air gap). These hotplates have the disadvantage of only keeping a relatively small area clear for vision. They also have a certain amount of aerodynamic drag, and might cost you a knot or so airspeed. On the other hand, they are usually removable without tools, and can be stowed when not needed. However, this sometimes leads to their riding along in the baggage compartment when they *are* needed.

A step up the sophistication scale is the integrally heated windshield with embedded fine wires. You'll find such installations on birds like the Navajo or King Air, for example. The network of resistance wires is almost incredibly fine. You have to lean very close to the windshield to see them at all. You may notice one small area of slightly thicker wires. This is a temperature sensor, and is used to control the windshield's electrical circuits to prevent overheating. Overheating a windshield can be serious business, particularly in a pressurized aircraft. You can make a quick and dirty check when the system's in use by putting the back of your hand against it. It should be perceptibly, but not uncomfortably, warm as felt from the cockpit interior.

Some fine-wire windshields may show very slight visual distortion when running at high heat, so it's not a bad idea to switch them off on final approach (unless you're in something horrible, like freezing rain). If the distortion ever appears more than usual, be on the alert for an overheated unit.

Another type of heated windshield is the conductive-layer type. Rather than using a network of fine wires, this type uses a single, very thin layer of conductive metal. Windshields of this type are easily recognized by a faint golden or coppery sheen in their appearance, like "disco" sunglasses, and their function and operation are virtually identical to those of the fine-wire type. The only significant difference is that the wire type operates on low-voltage DC, while the conductive-layer type requires AC supplied by an inverter, either a dedicated unit or the aircraft's AC system. Either type of windshield, when in operation, will cause very significant errors in standby magnetic compasses mounted atop the panel or on the windshield center post.

Yet another method is available in turbine-powered aircraft. In this method, hot bleed air is ducted through nozzles at the base of the windshield. Such systems are also fairly effective for rain removal, although operating them for any significant length of time at above-freezing temperatures may cause windshield overheating. Some provision is normally made to monitor bleed air temperature, often with automatic shutoff and annunciation in case the unit overheats. As with any bleed-air operated equipment, a slight performance penalty is paid as bleed air use increases.

Moving further aft, engine induction systems are the next candidates for ice protection. With fuel injection the rule, rather than the exception, for recip engines in this class of aircraft, carburetor icing, as such, has ceased to be the menace it once was. Other forms of induction icing can still occur, though, notably freezing up of moisture trapped in the induction air filter, and simple impact icing of the actual induction air scoop. In either case the solution is simple—alternate air, usually taken from some warm, sheltered spot within the cowling. We'll cover recip-engine induction systems in more detail in the next chapter.

Turbine Engines—Anti-Ice

Turbine engines operate a bit differently, and it's impractical to provide them with a separate alternate air source. Instead, provisions are made to protect them against icing in their regular induction systems.

Such protection, it should be noted, is definitely *anti*-ice rather than *de*-ice. This is because any ice, once formed at the engine inlet and then broken loose by some form of de-ice system, would most likely be ingested by the engine compressor section, causing problems ranging from flameout at best, to engine destruction at worst. In a sense, this is a turbine aircraft's most critical ice-protection area.

Incoming air is initially directed past the lip of the air intake, and this is the first point at which ice might form. Therefore, the intake lip is heated, either electrically, using a boot similar to a prop-blade anti-ice boot, or by running hot bleed air through a hollow metal section that forms the actual intake lip. Again, let me stress that such systems should be turned on *before* an icing encounter, not during it. (If an inadvertent icing encounter does occur, by the way, and you're flying a turbine ship, turn on the engine ignition *before* heating up the intakes: this will help prevent flameout if ice does break loose and get ingested.)

Intake lip protection is deemed sufficient for some engines; others, however, go one step further with an "inertial separator." This is a moveable vane which can be positioned by cockpit control in the air intake path, and it has the effect of forcing intake air to "go around a corner." Particles of ice and other moisture, being heavier, can't "make the corner" and end up "skidding out" into a different section of the intake path, where they're dumped overboard. Just as bleed-air systems possess a small performance penalty, so do inertial separators, since they don't allow the engine to breathe quite as freely as when they're not being used. We'll go into this in more detail in the chapter on turbine powerplants.

Airframe De-Icing

Airfame de-icing is the final ingredient, and again the name is important. The rubber boots used on recip and turboprop twins are strictly

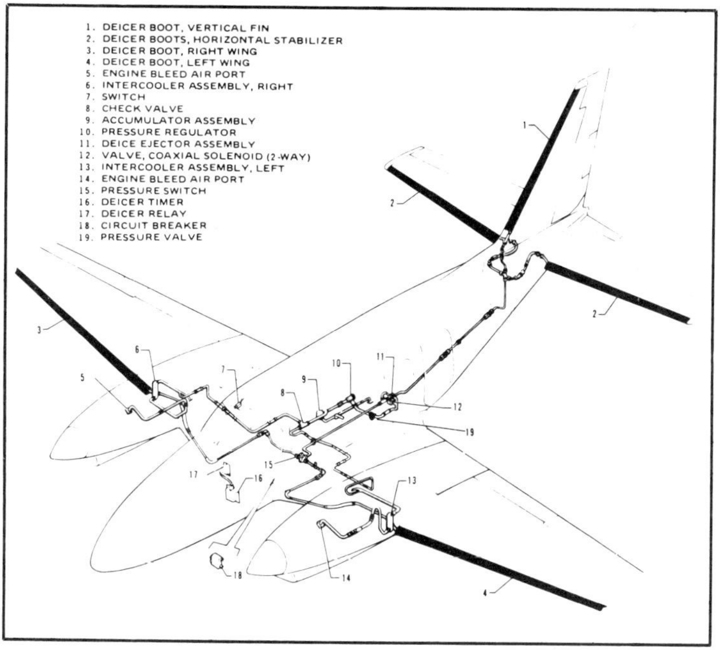

Fig. 3-16. A major use of pneumatic pressure in twins is for control and inflation of de-ice boots. This diagram of the Piper Cheyenne system shows the timer and solenoid valves that inflate different boots in the correct sequence.

de-icers (Fig. 3-16), and any attempt to use them in an anti-ice mode will not only be fruitless, but will render them useless as de-icers as well! Only pure jets, with their larger available quantities of bleed air, can afford to use it to heat entire wing and empennage leading edges. Propeller-driven aircraft generally still depend on the good old black rubber boots that have been with us since the 1930's.

The principle is a simple one: inflating the boots with air causes them to swell up, cracking off ice that has accumulated on them. This means, of course, that sufficient ice must be allowed to accumulate so that it will crack properly before activating the boots. If the boots are cycled too early, an "ice bridge" will form over them, beneath which the boots will pulsate uselessly up and down and upon which more and more ice will form as long as the icing conditions continues.

A typical system will include boots for the wings and empennage surfaces, an appropriate air source, and an inflation/deflation timer with a cockpit control switch. Each time the system is cycled, the timer applies air to either the wings or the tail. This allows sufficient time for the appropriate boots to reach full inflation, then it applies a vacuum to hold them flat against the aircraft surface while inflating the other half of the system as is needed. Inflating wings and tail sequentially, rather than at the same time, allows for more effective use of the air supply. Some sort of indicator or warning lamp is usually provided to allow the pilot to monitor boot inflation and deflation.

Condition of de-ice boots should be checked on preflight inspection. Minor scrapes and cuts can be patched like innertubes; major ones may require replacement of a boot section. Any air leaks, of course, impair the inflation and function of a de-ice boot.

Since they're non-conductive, boots tend to build up static electricity in flight. Most modern boots have an electrically conductive coating to reduce this effect. Check with the manufacturer's service manual before washing the airplane down, as some types of cleaner tend to scrub this coating off. Various spray-on products are also available which not only protect the boot and preserve its useful service life, but make ice crack off more easily, too.

Finally, as we continue aft, we'll find that some aircraft have special areas on their empennage that require ice protection. For example, narrow control-surface gaps that might be jammed by accumulations of ice. Such areas are usually heated electrically, either with boots similar to those on propeller blades or with electrically heated metal plates.

Environmental Control Systems

We'll conclude this chapter by touching briefly on some aspects of heating and air conditioning. Parts of this area will be covered in more detail when we discuss pressurized aircraft, since they employ specialized systems. Even the occupants of lowly non-pressurized aircraft, though, deserve a modicum of comfort, which means keeping them warm or keeping them cool.

Heating Systems

Unlike most singles, which use exhaust muffs for cabin heat, most recip twins have fuel-burning heaters. These are smaller versions of the oil burners that are used for home heating, and are *two-stream* devices. There's no direct flow of combustion-heated air into the cabin. Rather, the combustion air and fuel heat an inner tube in the heater, and air flowing around the outside of this tube is heated by it and ducted into the cabin. This prevents the chance of carbon monoxide or other undesirable combustion products from finding their way into the cabin and the lungs of its occupants.

In addition to the flow of air through them and into the cabin, combustion heaters require a minimum airflow through their burner portions. This is usually provided by a separate electric blower, although some units can get sufficient ram air to keep running once the aircraft is airborne. A larger blower forces outer-stream air through the heater and into the cabin, often shutting down automatically when the airplane is airborne and ram air can take over.

Additional components include a fuel pump, solenoid-operated fuel valve, a pressure switch which ensures adequate combustion airflow before turning on the fuel flow, and an ignition system with a coil, a vibrator, and a special spark plug or igniter.

Unlike, say, a burner on a stove, aircraft heaters cannot be "turned up or down." Instead, the cabin's temperature control sets a thermostat in the heater output which cycles it on and off to maintain a desired temperature in the duct.

Since a malfunctioning heater could be a potentially explosive situation, various safeguards are provided. I've already mentioned the pressure switch which cuts off the fuel flow unless there's enough airflow for complete combustion. There's also an overheat protection switch which shuts the heater down if the thermostat cannot regulate it to the appropriate temperature. Some of these overheat switches can only be reset on the ground.

Heater ignition systems have never been noted for their reliability. Therefore, several aircraft heaters have two ignition systems, or at least two sets of breaker points on their vibrators. The most failure prone component can be selected from the cockpit. If your trusty fire-breathing dragon fails to wake up on schedule, try the alternate set of points. (If, on the other hand, he wakes up with bad breath and nasty smells, shut him down, open all the vents, and don't try him again without a checkup.)

The situation is somewhat simpler in turbine aircraft, since there's always plenty of hot bleed air available. Since this air is tapped off upstream from the engine combustion chamber, it can be used as-is for the cabin. A mixing valve adjusts the proportion of bleed air and cold outside air to maintain the desired temperature.

Air-Conditioning Systems

A few years ago, air conditioning was considered a frill; currently it's quite common. We'll deal here only with the systems found on recips, both pressurized and otherwise. Turbine ships use special systems which will be covered when we discuss pressurization equipment. (Except the King Air, which uses a system similar to those discussed here.)

Any air-conditioning or refrigeration system is based on the physical law that when you compress something, it gets hotter, and if you allow it to expand, it gets cooler. Ever notice that tires show higher pressure when they're hot? In an air-conditioning system, a working fluid (a chemical that's particularly prone to these temperature/pressure effects) is compressed, making it warmer. The warm, high-pressure fluid is then ducted through a heat exchanger, rather like a car radiator, where it sheds its heat to the surrounding air. We now have cool fluid, still at fairly high pressure. If we allow it to expand, it'll become quite cold. Simple, isn't it?

In aircraft systems, the compressor is either driven by an engine (either hydraulically or by belts) or powered by an electric motor. The heat exchanger, or "condenser," is located somewhere conducive to cooling by airflow, usually in an engine nacelle (or behind those odd-looking louvers in the nose of a King Air). In a pressurized aircraft, these components are located outside the pressure vessel.

A second heat exchanger, or "evaporator," is located in the cabin air ducting at an appropriate point. This is where the expansion of the working fluid takes place, and the cold point in the system. Often, the same ductwork is used for both heating and cooling.

Some engine-driven compressors extract sufficient power from their respective engines so that they might adversely affect single-engine performance. Thus, many air-conditioning systems are placarded against use during takeoff and landing. This is not a real handicap. If you turn off the compressor as you take the active, the blowers will continue to run, and the system will have plenty of cool air stored up to run for half a minute or so with no noticeable change in temperature, and can be turned on again right after takeoff or landing. Air conditioning has been of particular value in pressurized airplanes, since the same laws of physics that make the air conditioner work dictate that the air provided by the airplane's pressurization system will be somewhat warmer than ambient air. In fact, many pressurized aircraft almost never have to use their heaters except at the highest altitudes and during the coldest outside air temperatures!

Fire Detection and Extinguishing Systems

Quite a few of the larger aircraft available now can be fitted, usually as an option, with fire detection and extinguishing (Fig. 3-17). This is both good and bad. It's good that such systems are available, and bad that they're options instead of standard equipment.

Detecting a fire can be done in various ways. One method uses heat sensors at various spots in the engine compartment. Another method uses infrared light sensors which are not inherently sensitive to heat, but will register the infrared radiation emitted by flames (or red-hot metal, for that matter). Some aircraft baggage compartments include smoke detectors similar to those used in homes, and some systems add these, too, to the nacelle detection systems. Whatever the system, when it's activated it'll light a large red light at some prominent location on the instrument panel. A self-test function is provided. The simplest ones just prove that the red light hasn't burned out, while the better ones check all the sensors as well.

Larger aircraft usually incorporated some sort of immediate firewall shutoff arrangement, either by punching the fire warning itself or an adjacent switch. This immediately cuts off fuel and hydraulic lines to the affected engine, and in some cases secures the generator on that side as well. In smaller aircraft, such tasks are performed one at a time by the pilot.

Now that we've detected a fire and secured the engine, how will we put it out? If the aircraft has a fire extinguishing system, there'll be a container of extinguishing agent in each nacelle. In some aircraft, it's plain

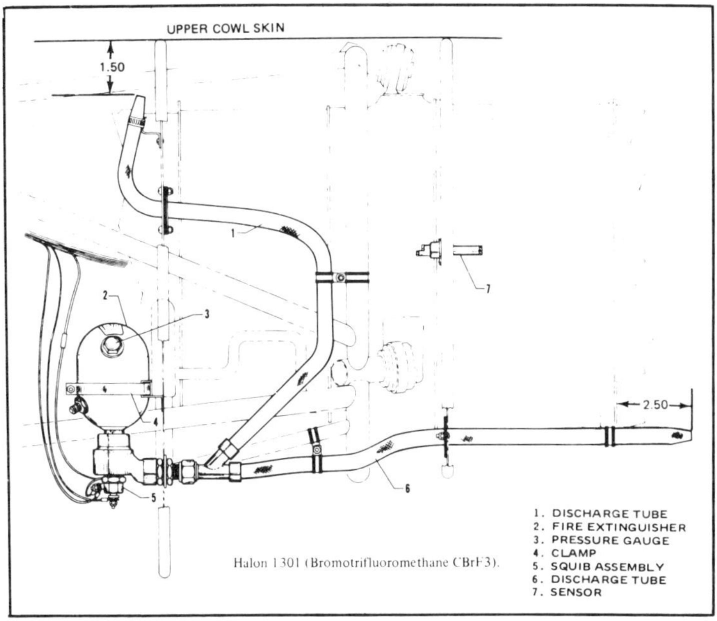

Fig. 3-17. A typical fire extinguisher system (this one is from a Piper Cheyenne) includes a pressure bottle of extinguishing agent in the engine nacelle and a means ("Squib") of firing it electrically; firing circuits are normally powered even when the aircraft electrical system is shut down. The pressure gauge on the bottle is a preflight check item. Some aircraft have provision for "crossfeeding" the extinguisher, allowing both bottles to be used in succession on a single engine.

old CO_2. Other systems use bromine derivatives, dry chemicals, or a Freon gas derivative called *halon.* If you haven't already secured pressurized air from the affected engine to prevent smoke and fumes from entering the cabin, be sure you do it before triggering a bromine fire extinguisher, because the stuff's poisonous. Fire bottles are usually fired electrically, and their controls can include arming switches (sometimes linked with the firewall cutoff switch) and a test button so you can be sure they'll work when needed.

The niftiest news in the fire-extinguishing world is the Halon system. Not only does it do a good job putting out fires, it'll snuff out cabin fires even if it's at low enough concentration to be breathed with no adverse effects. A number of manufacturers now offer systems that will flood the cabin with Halon at the press of a button; if you don't have a system like that, you can buy portable Halon fire extinguishers, and they're certainly safer and easier to use than the dry-chemical type. If nothing else, you can quietly crack the valve when you see the chairman of the board in back preparing to fire up a noxious cigar, with Halon he won't even be able to light a match!

4

Introduction To High-Performance Reciprocating Engines

The reciprocating engine has come a long way since the big, oily radials that powered corporate and executive ships as recently as 20 years ago. Today's small, light opposed engines operate at levels of efficiency that would have been unheard-of in the past, and are one of the main reasons that the pressurized recip twin remains the most fuel-efficient corporate or executive aircraft.

Of course, the rules that govern the basic operation of any four-stroke gasoline-burning recip haven't changed since the Wright Brothers, and the basic *core engine* of even the most sophisticated modern recip powerplant isn't all that different from the little four-banger in a J-3 Cub (or, for that matter, in a Volkswagen). It's the various gadgets and "gilhickies" that have been hung on the engine over the years that have made it more complex, more expensive to operate and maintain, and less forgiving when improperly operated, but at the same time allowing it to produce power on the order of one-half horsepower or better per pound of engine weight. That's better than twice as efficient as some of the older engines, and the improvement in specific fuel consumption, that is, how much fuel must be burned to produce a given amount of power, has been similarly dramatic.

Recip aircraft engines can be direct-drive or geared. They can have carburetors or fuel injection, and they can be normally aspirated, mechanically supercharged, or turbo-charged. Each of these factors is represented in the by-now standard nomenclature and abbreviation, together with engine displacement; thus, a look at the specs of an airplane with, say, GTSI0-520-L engines reveals to the informed eye that each engine is a *G*eared, *T*urbo*S*upercharged, fuel-*I*jected, *O*ppossed (i.e. flat powerplant with a displacement of *520* cubic inches, while the -L designation represents a particular model and accessory configuration of this powerplant. (Incidentally, for some reason known only to them, mechanics pronounce the designation for such engines "gitso.") Once again, we'll look at specific engine types and take a system-by-system tour through the inside of a typical powerplant installation.

Geared Vs. Direct-Drive Engines:

Geared opposed engines may slowly be coming to the end of their heyday, as some direct-drive types begin to approach their performance while offering lighter weight and less maintenance. Nonetheless, the larger geared mills will be around for some time. Why have geared engines in the first place? For more power from a given engine size, as I'll elaborate upon.

The measure of how much power an engine, *any* engine, can produce is related basically to a very simple factor—how much fuel and air can it burn in a given time? Obviously, bigger engines can burn more at a time than little ones. A turbine, although small in size can shove immense amounts of fuel and air through itself in a short time, hence its high power-to-weight ratio.

Of course, you can't just increase horsepower indefinitely by making your engines bigger and bigger. After all, the airplane has to be able to lift itself, and some reasonable payload, off the ground. How can we get more power out of a small engine? By burning more fuel and air in a given time, in other words, by running it *faster*.

Of course, this simple solution has its limits. There comes a point at which more engine speed creates such friction losses in the internal part of the engine that all its power goes just to keep itself running. Long before that though, we reach another critical point, at which the rotational speed is simply too fast for efficient propeller operation, and *that*, is why we have geared engines! For a reasonable-size propeller, say eight feet or less, somewhere between 2500 and 3000 rpm, is where efficiency really starts to sag. As the prop spins faster, the tips get closer and closer to Mach 1, the speed of sound, and their drag rises tremendously (so does the noise!). Geared engines offer a way of running the core engine at higher, more weight-efficient speeds while holding prop rpm down to reasonable values.

Why aren't all engines geared? Because you pay a penalty. For one thing, the gears themselves weigh quite a bit, cost an even larger bit, and tend to be maintenance-prone. Also, geared engines are very sensitive to improper operation, particularly in two specific areas:

Back-driving. This isn't good for any aircraft engine, but it's particularly bad for geared ones. We are referring to situations in which the aircraft is descending or slowing, and the propeller is driving the engine, rather than vice versa. All the power-producing and transmitting components of the engine, including the pistons, connecting rods, large and small rod bearings, crankshaft bearings, and gearing system (if any), are designed to *make* power and shove it out the main shaft. They are not to be driven back and forth; excessive back-driving speeds can cause bearing wear and can lead to rod and piston slap, damaged crankshaft counterweights, and even prematurely broken crankshafts. Since the internal parts of a geared engine turn proportionately faster than those of a direct-drive type, the problem is magnified. While pilots are always cautioned not to operate engines at too *oversquare* a power setting (at high manifold pressured and low rpm) those flying behind a pair of gear-gnashers should be equally cautious about flying at too *undersquare* a setting. Descent and approach planning in a geared-engine airplane can be a good deal more critical than that of a direct-drive ship, which can be a real hassle in some ATC environments (particularly near large terminals where the FAA has a "keep 'em high" policy).

Lugging. This is another problem to which geared engines are prone to experience, particularly when they're operated by pilots who are more used to the direct-drive types. You may recall the word "lugging" from when you learned to drive a car with a stick shift. It was that loss-of-power condition you experienced to climb a hill in high gear. In an airplane, it is that same "condition" you obtain when you're running too "over-square," while carrying too much manifold pressure at too low an rpm.

Don't Get Psyched Out!

What's ironic about this problem is that its cause is largely psychological. Pilots used to direct-drive engines are just plain scared of those big numbers on the ol' tachometer. For example, an engine with a gear ratio of 1.67 (which sticks in my mind as the gearing of some engine I flew at some point, could it have been the old 680 Commander?) will be turning 4000 rpm on takeoff for a prop rpm of only 2400! Those are heady numbers indeed, especially to someone who's already dumb-founded by high manifold pressures common on larger super-charged engines. The result is improper operation, with the pilot often setting far too low a prop control position, with consequent engine lugging, detonation, burned valves and pistons, warped rods and cranks and even cracked cases in extreme situations. There are, of course, at least two solutions to this problem. One is a simple matter of training and education—fly the engine by the book, and you'll be fine. An even simpler aid is one used by Cessna—they just calibrate the tach in prop rpm. I'll bet not one 421 driver in 10 really knows how fast his crankshafts are spinning!

The Continental Tiara Series

This might be a good time to mention one particular type of geared engine that never quite made it—the Continental Tiara series. One weakness of most geared engines is that their gearing design was something of an afterthought. The basic engine is usually a direct-drive type developed originally for lower power, with the prop gearbox more or less bolted on later. This is another reason that many of them tend to have short, unhappy lives. For example, the geared 520-inch engine found in the Commander 685 or the experimental Bellanca Skyrocket is squeezing 435 ponies out of the same basic powerhead that has to work to produce a paltry 285 hp in the Bonanza or Cessna Centurion!

The Tiara, on the other hand, was designed from the start as a geared engine. In fact it saved weight and complexity by using a 2:1 gear ratio and no separate reduction gear. Instead, the camshaft was driven by a pair of the biggest, ruggedest timing gears you've ever seen, and it provided output drive as well. The crankshaft was specially designed for high-speed operation, with all kinds of nifty torsional dampers. The whole engine was engineered for efficiency, as well as relatively low-cost production on automated equipment. A giant factory was set up in Alabama to make them.

The Tiara was, unfortunately, a failure. Its efficiency was no better than that of other geared engines; neither were its reliability or its price. Only one model was developed (a whole lower range had been planned), and as of this writing, some 10 years, I know of only three aircraft in which it's used: an agplane, a European 4-place, and a homebuilt amphibian!

A discussion of geared engines and their inefficiency leads naturally into an area of similar inefficiency and maintenance-proneness: mechanical supercharging. By now, turbocharging has become so common that younger pilots can easily forget that it wasn't always that way, superchargers used to be run by gears attached to the crankshaft, in both radial and opposed engines.

The GSO and IGSO engines are slowly disappearing from the scene, partly because they've become so expensive to maintain that it's often cheaper to replace them altogether with more modern engines as they come up for overhaul, but I'm sure that some of them will fly on for a few more years. You're most likely to find them in the heavy, unpressurized twins of a few years ago—Beech Queen Airs, 680-series Commanders, and the like. In these engines, a supercharger impeller is mounted on the back of the engine and spun at rather remarkable speeds (20,000 rpm in one type) by a complex gear train coming off the crankshaft. Needless to say, these super-chargers and their associated drive systems love maintenance, and even when well-maintained they can pack it in suddenly and spectacularly. Slow, smooth throttle manipulations and *strict* adherence to engine operating limits are the watchwords here.

Moreover, the geared superchargers are quite inefficient. The amount of boost they can provide depends directly on engine speed, and is limited in any case. The harder you work them, the more power they require. At altitude, with the blower on, an IGSO-380 at cruise might be using up a third of the engine output! Even then, there's not enough bleed air to spare for anything but the engine itself. The early pressurized aircraft, such as the Commander 720, required a separate cabin supercharger mounted in the aft fuselage and powered by an evil 3000-psi hydraulic system. It was easy to tell 720 pilots in the bad old days; they were the guys with red stains on their shirt cuffs (from hydraulic leaks) and a bottle of "Super-Anahist" in their shirt pockets to keep their sinuses clear.

Modern Turbocharging Systems

If the mechanical supercharger was a problem, the turbocharger is just the opposite—turbos are light, compact, relatively inexpensive, simple to maintain, and *efficient*. In fact, it's the development of the turbocharger to which we owe the existence of most of the pressurized light and medium twins we fly. With mechanical blowers, they just wouldn't have been economically or operationally feasible.

We can even interject a patriotic note—the original development of the turbocharger, in World War II, not only was a tremendous strategic advantage (it made possible such aircraft as the B-17, the B-24, the B-29, and the P-38), but the advanced metallurgy required to make turbos that wouldn't melt at their normal operating temperatures led directly to the development of the first parctical jet engines. You may not realize it, but if you could look into your cowlings while running at typical cruise power, you'd see your entire turbocharger glowing somewhere between cherry red and orange!

What's involved in a turbocharger system? Relatively little, considering all that it can do for you. At its heart is, of course, the turbocharger itself (Fig. 4-1).

Compared to a mechanical supercharger, with its gear trains and clutches, a turbocharger is the eptiome of simplicity. It has exactly *one* moving part—a shaft with two wheels, one at each end, enclosed in a pair of specially-shaped chambers and supported by bearings between them (Fig. 4-2).

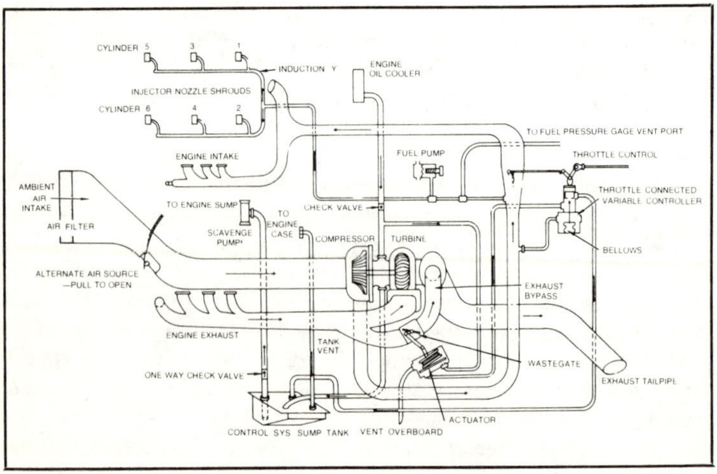

Fig. 4-1. Schematic of a typical turbocharger system (courtesy of FAA).

Fig. 4-2. This cutaway Cessna 421 powerplant shows the location of the turbocharger at the rear of the engine. Note the short exhaust connector pipe bypassing the turbo assembly; this is the location of the wastegate. The round object on the side of the engine ahead of the #1 cylinder is the alternator.

One side of the turbocharger is very similar to a mechanical supercharger—this is the "cold," or induction, side, and the wheel on this end of the shaft is called an *impeller*. It's really a centrifugal pump for air. Induction air is taken in at the hub of the impeller, and "slung" toward the periphery by centrifugal force. While this force may not seem too significant when you consider something as insubstantial as air, remember that the impeller is spinning at speeds approaching 100,000 rpm! Now compressed to high pressure, the air is picked up around the edges of the impeller and ducted off to the rest of the engine induction system.

The other, or turbine, end of the turbosupercharger is the "hot" or exhaust side. This is the portion of the unit that gives it both its name, *turbo*supercharger, and its efficiency, since it recovers energy from the engine exhaust that would otherwise simply be lost overboard. The turbine wheel and its enclosure are, in effect, a mirror image of the impeller on the cold side, and work in exactly the opposite fashion—hot, high-pressure exhaust air is ducted to the *edges* of the turbine wheel, transferring its energy to the wheel and spinning both it and impeller at the other end of the shaft. Then, with its work done, the exhaust is ducted away from the center of the turbine and dumped overboard. It's almost as if you're getting something for nothing, since you're using exhaust energy that would otherwise go to waste. Units of this sort were once called "bootstrap" systems, since it's almost like picking yourself up by your own bootstraps, and the name persists for certain types of turbine operation which we'll discuss in a moment.

The most basic turbocharging system, then, would consist of nothing more than a turbocharger with its cold side plumbed into the engine induction system and its hot side plumbed into the exhaust system. Such a system would, in fact, work, and some hombuilt aircraft types make do with little more. For easier and more stable operation, though, and better controllability with reduced pilot workload, most cabin-class and pressurized twins use slightly more elaborate systems.

Operating a Turbocharger

Let's consider the operation of a simple system like the one I just described, so that we can see its problems. We will start with the airplane idling at the end of the runway. The engine is at idle rpm, and the turbocharger is spinning away at some arbitrary figure. Now we'll open the throttle for take-off. As we do, the turbocharger may initially slow down for an instant, since it now has to pump a larger volume of induction air. In a moment, though, the increased exhaust flow will have brought it back up to speed. In fact, it'll now be producing *too much* induction air, leading to even *more* exhaust flow, and leading to even *more* induction air yet. You can see that there will be a certain amount of lag, and that it'll take a fine hand on the throttles to avoid walking all over the manifold-pressure redlines. The turbine is operating in what we'd call *bootstrap* mode. Any change in one operating parameter will be amplified and will appear somewhere else as well.

All right, we've gotten off the ground without blowing anything and it's time to make the first power reduction. Cautiously, we pull back on the throttles until the MAP indicates the correct value. Now, let's bring back the props. Got 'em synched? Good . . . hey what happened to the manifold pressure? Well, when we set the props for climb, we reduced rpm. This, of course, means that less air flowed through the engine, so, naturally, the turbo slowed down, and, as it did, it pumped even *less* air through the engine. The same thing will happen again when we make the next power reduction for cruise. How about mixing adjustments? When you lean the mixture, you're reducing the *mass* of the airflow though the engine, since there's less fuel mixed with it, the blower will slow down. Do you change airspeed? Sure, but as it increases, more ram air will be supplied to the blower, which will thus increase its output, as will the aircraft engine, so the airspeed will try to increase some more. Do you see now why it's called the bootstrap mode?

Actually, the above horror story isn't just to make you appreciate the niceties of an automatic turbo control system. Automatic systems work only because they offer a certain amount of superfluous turbocharger capacity which is controlled by a device called a *wastegate*. At some particular altitude, which varies from type to type, the auto control system will have come to its limits and the wastegate will be completely closed. From that altitude on, continued operation is highly feasible but it will be in "bootstrap" mode. When you fly a turbo-charged airplane, be aware of its "critical altitude," and realize that "bootstrap" behavior above that point isn't a malfunction, it's just one of the facts of life.

An automatic control system incorporates the turbo-charger itself and a control device called a wastegate. The engine exhaust system "forks" upstream from the turbine, and the wastegate can close off the leg that bypasses the turbine. With the gate open, all exhaust flows overboard around the turbocharger, which produces no output. With the wastegate closed, all exhaust flows through the turbocharger, which spins at maximum output. Intermediate positions of the wastegate can regulate turbocharger speed. As the aircraft climbs higher and higher, the wastegate will be driven further and further toward the closed position by the control system. At critical altitude, the wastegate is completely closed, and bootstrapping begins.

The wastegate is operated by engine bleed oil pressure controlled by a *differential controller.* This pressure-sensitive unit is connected to the intake system upstream from the throttle plate, as well as having a pressure reference connected to the intake manifold downstream from the throttle. (These pressures are referred to as upper and lower deck pressures, respectively.) The differential controller regulates turbocharger output (by controlling the oil pressure sent to the wastegate actuator) to provide a constant manifold pressure as well as a constant difference between upper and lower deck pressures. Thus, throttle position will remain unchanged during a climb, while overall efficiency is improved running the required manifold pressure). Generally, upper-deck pressure is set a few inches higher than the desired manifold pressure. In pressurized aircraft, this difference tends to be set higher, so that adequate bleed air is available for cabin pressurization, even when power has been reduced for a descent to lower altitudes.

There's an additional controller in the system. It is called an *absolute controller*, and its purpose is to prevent the system from producing excessive manifold pressure or turbocharger output. It also prevents turbocharger overspeed. Normally inactive until the turbocharger output approaches the limit, it then dumps a protion of the wastegate oil pressure, thus opening the gate and reducing turbocharger speed.

Some simpler, lower-cost systems dispense with active controllers, using instead a fixed ground-adjustable wastegate or a mechanical wastegate linked by cams and rods to the throttle control. These units offer adequate control, but are always in a sort of "modified boostrap" mode, since they lack active automatic controls. In these systems, overboost protection is usually provided by a "pop-off" valve in the induction plumbing which simply opens and dumps excess induction air overboard at a preset pressure.

If the aircraft uses turbocharger output air for pressurization, it's bled off the upper deck before use in the induction system. A special venturi reduces its flow rate and temperature before it's ducted to the cabin.

Turbocharger output air is generally quite warm. Not only has it been heated by compression, but the turbocharger unit itself is physically quite hot due to the exhaust flow through the turbine side. This heating reduces the density of the output air. Thus, a manifold pressure of 30 inches in a turbocharged aircraft at some altitude would not produce as much horsepower as the same manifold pressure in a normally aspirated engine at sea level. Manufacturer's power charts reflect this difference. The higher the altitude, the greater the compression (and hence the higher the temperature of blower output air), leading to typical manifold pressure increases to around .2-3 in. Hg per 1000 feet. Some aircraft incorporates air-to-air heat exchangers, or "intercoolers," between the turbocharger and the induction system to minimize these effects and increase the aircraft's critical altitude.

One way of accurately setting power at altitude would be to adjust each engine for optimum mixture (using EGT), set rpm, and then use the throttles to match fuel flow. Unfortunately, most fuel flow meters on the class of aircraft in question aren't quite accurate enough, so we'll generally have to stick with good old manifold pressure and rpm. Accurate power settings require a correction for outside air temperature, and in the faster airplanes, there may be a further correction between indicated and true outside air temperature.

We'll get back to operating techniques in a bit, let's now take a quick look at the rest of the engine's systems.

Oil Systems

Most of the engines in use today have simple, selfcontained oil systems in which the crankcase is used for storing the oil. Some, however, have *dry sump* systems in which an external tank is used, while *scavenge pumps* pick up used oil at specific points in the engine and return it to the tank. Otherwise, there's no difference in principles of operation.

It should be kept in mind that, in an aircooled engine, oil does more than just lubricate the moving parts—it has an important cooling function as well. Thus, most engines are equipped with an *oil cooler*, and heat exchanger in which the oil can pass its extra heat on to the passing air. The flow of oil through the oil cooler is generally controlled by a thermostat which automatically bypasses the heat exchanger if the oil is below a certain temperature.

In very cold weather, oil left in the oil cooler after shutdown can congeal into a rubbery mass which will continue to block the oil cooler even after the bypass valve opens. This leads to engine oil overheating, with consequent breakdown of its lubricating qualities and loss of oil pressure. Some types of oil coolers are less prone to this than others. Also, some aircraft have *oil cooler winterization plates* that can, and should, be installed if low ambient temperatures are expected. The old radial engines were particularly sensitive to correct oil temperatures, and generally offered the pilot a host of controls, including manual bypass valves and oil cooler shutters, to help regulate oil temperatures.

Additional oil system components include an oil filter, nowadays often a disposable-element type, and temperature and pressure gauges.

The gauge generally indicates pressure and temperature at the oil pump output, before the oil cooler. The importance of regular filter and oil changes in a high-performance engine cannot be overstressed. It's good practice, each time a filter is changed, to indulge in the rather messy practice of ripping the discarded filter open to check for metal particles. Even better is a so-called *soap* check. This stands for Spectroscopic *O*il *A*nalysis *P*rogram. At regular intervals, a small oil sample is taken and burned in a spectroscope, which reveals if any trace metals are present. Proper interpretation of SOAP results can pinpoint wear of critical engine components long before any overt symptoms occur.

Induction Air Systems

Most turbocharged aircraft offer not one, but two alternate air induction systems. In normal operation, ram air is run through a filter to eliminate contamination, then it is fed to the intake side of the turbocharger. Should the filter become blocked, a manual or automatic induction door can be opened downstream of the filter to allow continued operation. In many aircraft, this alternate source is taken from some sheltered spot within the cowling, and is hence more or less free from icing considerations. Some manufacturers judge that air taken from the cowling could be potentially contaminated, and require that pressurization source air from a given engine be shut off when the engine is operating on alternate induction air.

Most systems include a second, automatic alternate-air door in the induction system *downstream* from the turbocharger. Normally, a turbo failure has no effect on induction system performance other than the obvious reduction in manifold pressure, since there's plenty of room for air to be sucked through the impeller even if it's not moving. It's conceivable, though, that a catastrophic turbocharger failure could block the plumbing with pieces of shattered impeller or the like, so the secondary door is rigged to "suck open" at a given pressure differential. Usually, a magnetic latch is used, so that once open the secondary door will remain open until closed manually, or until turbocharger operation resumes.

Ignition Systems

Aircraft engines are almost universally fitted with dual redundant ignition systems powered by magnetoes independent of the aircraft electrical system. Some recent engines use a *dual magneto*, consisting of two separate systems sharing a single physical case (and a single drive gear, if that goes, so does the engine).

Aircraft destined for high-altitude operation often have special ignition systems. This is because the thin air at high altitude presents less resistance to arcing (sparks) and corona discharges (eerie blue static electricity discharges) than the denser air near sea level. High-altitude magnetos use special insulation or pressurization to reduce the problem, and they are used in conjunction with heavier spark plug harnesses.

Many high-performance engines also have special ignition provisions to make starting easier. These can include automatic retardation of one magneto while the starter is engaged, and a starting vibrator (tradenamed "Shower of Sparks" by one manufacturer) which, as the name implies, increases spark duration for starting.

Fuel Injection Systems

In high-performance engines, fuel injection is the rule. Some older types use *injection carburetors*, which resemble carburetors in appearance but which are, in effect, single-point fuel injection systems. Most, however, use individual injection nozzles for each cylinder.

Fuel injection requires relatively high fuel pressures, higher than those that can be supplied by a gravity feed fuel system. Thus, even high-wing aircraft, such as Aero Commanders, have engine-drive fuel pumps, which accept fuel feed from the tanks or the boost pumps and raise it to a pressure suitable for injection into the intake manifolds near or at each cylinder intake valve. Some fuel pumps are variable displacement types and are connected via mechanical linkage to the throttle and/or fuel mixture controls.

From the pump, fuel flows to the fuel injector servo (or "body"). This unit, which also contains the actual throttle plate, is a sort of hydromechanical computer which measures throttle opening and/or airflow, and then meters fuel accordingly. A mixture control varies the metering ratio to offer the pilot control over the fuel mixture.

Much of the fuel pumped to the metering system by the high-pressure mechanical pump is excessive for average power requirements, and it is returned to the aircraft fuel tanks. The remainder, the actual metered flow that runs the engine, is piped to a flow divider, or *spider*, which distributes it equally among the fuel injector nozzles. Aircraft fuel injections operate on a *continuous flow* principle, in which fuel flows through the system whenever there's adequate pressure and the throttle and mixture controls are set for flight. During the brief period when the intake valve is not open, fuel from its associated injector *pools* in the intake manifold, and is sucked into the cylinder as the valve opens.

Although some fuel-injected engines are fitted with primers for extreme cold-weather operation, most do not require them. Instead, the electric boost pump is run briefly before starting an engine in order to supply an initial charge of fuel to all cylinders. The characteristics of the continuous flow system make it possible to overprime or flood an engine.

Fuel boost pumps are provided in the aircraft fuel system, both to help move fuel from the tanks to the engines and as a backup in case the engine-driven pump fails. Many electric boost pumps are not intended for long-term continuous operation. If an engine fails due to mechanical pump operation and is restarted with the electric pump, a precautionary landing as soon as possible is often an excellent course of action.

A word of caution should be interjected about the fuel flow gauges found in most recip twins. In all but a few cases, these are not true fuel-flow gauges; rather, they indicate fuel-injection pressure. If all is functioning properly, this can indeed provide a good approximation of fuel flow. However, if an injector becomes clogged or blocked, back pressure in the system will rise and the fuel flow indicator will indicate an erroneously high rate of fuel flow. If the pilot now leans the mixture to a precomputed fuel flow indication, the actual flow will be insufficient, and the engine will be running overly lean. Therefore, if you see an abnormally high fuel flow indication, check for signs of a blocked injector (e.g. engine roughness) and use other instruments as a guide for leaning. Sometimes, cycling the mixture control from full rich to a leaner position a few times will produce enough of a pressure surge in the injection lines to clear the blockage.

Exhaust-Gas Temperature Indicators

While we're on the subject of fuel injection and leaning, let's touch for a moment on one of the most useful engine instruments—the EGT, or *E*xhaust-*G*as *T*emperature gauge. Depending on its placement in the exhaust system of a turbocharged aircraft, the instrument may also be called a *TIT* (Turbine-Inelt Temperature) gauge. Either way, no high-performance aircraft should be without one for each engine (Fig. 4-3).

Fig. 4-3. Preparing a recip for installation in a cabin class twin at the Cessna factory (courtesy of Cessna).

The most common use for the EGT, of course, is in setting optimum fuel mixtures. The ideal mixture, called a stoichiometric mixture, is one in which there's just enough fuel to be completely burned in the charge of air available in the cyliner. This is the most efficient way to burn fuel, and is reflected by the highest (peak) exhaust gas temperature. Any richer, and the exhaust is cooled by unburned fuel; any leaner, and it's cooled by unburned air. Why don't we run our engines at peak EGT all the time? Because for high-power operations, we need to use some unburned fuel for additional cooling. Most engines can be cruised at 65 to 75% power at or near peak EGT, however, and it's a more accurate method of mixture control than most fuel flow indicators. Turbine aircraft operate under an additional constraint, in that peak EGT might be higher than the maximum permissible TIT. Operation above that value can cause damage to the turbine and bearings.

An even handier use, at times, for the EGT is as a method of pinning down isolated problems. Often, the savings in maintenance in a single occurance can pay for the cost of changing an EGT into an *analyzer* by installing a selector switch and a temperature probe for each cylinder. For instance, let's take the blocked-injector scenario mentioned above: You see a high fuel flow indication on one engine, so you switch the EGT from its TIT position to each cylinder in turn. If you're already operating at peak EGT, all the cylinders should show about the same EGT except for the blocked one, which will show a lower temperature. If you're operating on the rich side of peak, the blocked cylinder will either show hotter (if partially blocked, and hence too lean) or cold (if completely blocked, and not firing at all). Similar techniques can be used to isolate fouled plugs or sticking valves, and even if it's just a plug, it can save a lot of time when you taxi up to the mechanic and instead of saying, "Check the plugs on the left engine," you say "Change the bottom plug on #3 cylinder on the left engine."

Engine Cooling Considerations

The more sophisticated an air-cooled engine is, the warmer it seems to want to run. This is particularly evident in some of the newer planes, which have very tight cowlings. An engine cooling system consists of the cowling itself, the various internal baffles that direct air over the engine and between its cooling fins, and often cowl flaps or exhaust augmentors to aid cooling over a wide range of airspeeds.

Exhaust augmentors are one of the simplest and most elegant means of ensuring good cooling, although some of the larger engines now in use simply require more variable cooling capacity than augmentors can offer. Exhaust augmentors are the large, straight exhaust tubes, often chrome plated, seen protruding rearward from the nacelles of such ships as older "round nacelle" Commanders, Queen Airs, some 400-series Cessnas, and even some airliners such as the Convair 440. They work on the same principle as the venturi pumps mentioned in the section on fuel systems, since they're open at their forward, or nacelle, end as well as at the back. The actual engine exhaust pipe protrudes a short distance into the augmentor tube. High-speed exhaust gas discharged into the tube entrains a large amount of cooler air from the cowling interior as it goes. What's nice about augmentors is that they require no attention from the pilot. The more power you pull, the better they work. Unfortunately, they're also rather draggy when compared with a (closed) cowl flap, and sometimes they don't provide quite as much cooling at low airspeeds.

Cowl flaps regulate the exit area available for cooling airflow, and they are usually positioned either mechanically or electrically. On some aircraft, they can be seen from the cockpit, and are adjusted by eye; on others, an indicator is provided. Still other airplanes offer only full-open and full-closed positions, or leave the pilot to monitor oil and cylinder head temperatures and guess cowl flap position from them.

For some reason, not too much attention is paid in most training programs to the effect cowl flaps can have on aircraft performance, especially large ones. For example, the cowl flaps on the Piper Navajo Chieftain can knock 15 knots off the cruise if left all the way open! At lower airspeeds, their effect is proportionately less, so it's unlikely that retracting or failing to retract a cowl flap will make or break your day if you have an engine failure at low altitude. A possible exception to this might be found in the flat-nacelle Aero Commanders, which have a pair of good size cowl claps atop each nacelle, ideally positioned to serve as spoilers. I used to demonstrate their effectiveness to students by using the differential cowl flaps to make climbs, descents, and standard rate turns in both directions without ever touching the yoke or any of the trim wheels or power controls!

High Performance Engine Operation Tips

Operating a high-performance engine is no different from flying any other; it just requires a bit more caution and planning. We'll assume that we're dealing with a direct-drive, turbocharged, fuel-injected engine cooled by cowl flaps.

Starting

For some reason, a lot of big fuel-injected engines are easier to start than the little ones. Although this may sound like overgeneralization, most of them are very nice when cold, but rather less so when hot, at which point they're easy to flood. We'll try a cold start first.

The idea here is to get sufficient fuel into the cylinders for a gentle light-off, but not so much that we're overprimed. In addition to increasing the risk of backfiring or even a induction fire, overpriming drenches the cylinder walls with raw fuel, washing away the oil film and causing tremendous increased wear and scoring during the first few seconds of operation. A little care right at the outset can add hours before it's time for that expensive top or major overhaul.

Every manufacturer has their own recommendations, and every pilot his or her own pet technique, but in the end they all boil down to the same requirements—the right amount of gas, the right amount of air, and a starter system that can crank the whole "shebang" around fast enough for it to fire up. If it's at all cold, it's a good idea to make 100% sure that magnetos are off and mixture at idle cutoff, then go out and pull the prop through twice as many turns as the engine has cylinders (12 turns for a 6-shooter). This *frees up* components and layers of oil, and makes it much easier for the starter to do its thing.

With the *before starting engines* checklist complete, there's not much left to do but prime and crank. Ideally, the throttle should be set only very slightly above idle, so that the engine "takes itself away" from the starter as gently as possible, rather than starting it up with a roar. Those first few seconds of operation can be as hard on the engines as all the rest of the day's flying. Once it's running, reduce the rpm if possible and monitor the oil pressure gauge to ensure proper pressure rise.

What about when it's hot? The tendency is usually to overprime, which leads to flooding. If the engine's been run within the last hour or so, it's often sufficient to merely crank it with the mixture at ICO. There'll be enough residual fuel vapor in the cylinders for a smooth lightoff, and you then smoothly enrich the mixture. Often, this operation requires a boost pump running in "low" for a couple of minutes after the start to help purge fuel vapor from the lines.

What about those times, usually when the engine's hot, when nothing seems to work? Often, the problem is that you've had a couple of attempts during which the engine almost caught then died again, and you're not sure whether it's flooded or too lean.

It's at times like these when it may become necessary to intentionally overprime the engine; at least that way you know where you stand. Please note that this is *not* normally recommended procedure. It's hard on the engine, hard on the starter, and you are risking an intake manifold fire, but it usually *will* start the engine. It's a good idea to have someone standing by with a fire bottle, and if an APU is available, use it by all means.

First, let's flood the engine: boost pump on, mixture rich, throttle open halfway or so, and wait for a healthy indication on the fuel indicator. Now, bring the mixture back to ICO. Leave the throttle where it is, and leave the boost pump on for vapor purging.

Clear the propeller area and start cranking. You may even get a couple of weak coughs, with black smoke from the exhaust, but it won't catch. Keep cranking. What we're doing here is bringing in more and more air to the over-rich cylinders. With each groan of the starter, the mixture is coming a little closer to the correct one. When it reaches it, the engine will start, and pretty vigorously, too. *Slowly* advance the mixture while retarding the throttle; if you do it right, the engine will never get going all that fast before you get it under control. Leave the boost pumps on for a couple of minutes.

What if you get an induction fire? Don't panic, at least not yet. Keep cranking. Remember, that engine's about to start, and when it does it'll suck the fire down into the cylinders where it belongs. You've been cranking for 30 seconds and the flames are getting bigger? Ok, now you can panic or better yet signal the guy with the CO_2 bottle to move in and give it a shot. Let things cool down, inspect for damage, and try again.

Induction fires sound horrifying, but they seldom cause any damage. I used to fly a Cessna T-50 "Bamboo Bomber" whose two mighty Jacobs radials were a pair of the greatest little fire producing engines I've ever seen, and I never got an induction fire that lasted more than 10 seconds or failed to disappear instantly as soon as the engine started. A friend of mine, an airline type, related that when the first Super Connies with turbo-compound engines appeared, there was a standing offer of a bottle of Scotch to the flight engineer who could start all four engines without setting at least one afire, and the offer stood for several weeks before someone finally claimed it by cheating (he started the engines during a torrential rain).

What about when it's really cold? Here, the tendency is to under-prime, and it's also a time when batteries and starters aren't exactly at their best. An APU may sound like the sissy way out, but it'll sure help. (Incidentally, when you're using an APU in either hot or cold conditions, the temptation is strong to just keep cranking away until the poor engine just starts from boredom. That's a great way to melt a starter; 15 seconds should be the maximum, with a good minute and half cool down period. If you've flooded an engine on a hot start, those 90 seconds will be just about right to let the vapors dissipate, too.) When it's cold, give the engine plenty of prime, don't hesitate too long in advancing the mixture when it fires, and be ready to come in with a few shots of high boost pump if necessary to keep things going for the first few seconds. Running a really cold engine at any sort of power is terribly rough on it, so don't do it. Just sit there for a couple of minutes waiting for the oil pressure to stabilize, and it may well be considerably higher than normal because the oil's so gooey, so enjoy the cabin heater, copy your clearance get out all your maps and charts, and give the poor engine a chance to get at least some of its metal parts to similar temperatures.

This doesn't mean, however, that you should spend too much time on the ground, even when it's cold. Our relatively small opposed engines don't need as much of a warmup as the old radials, and excessive ground running will produce localized hot spots in the engines that contribute to thermal stress and a shortened overhaul life. We hear various criteria mentioned for determination of when an engine is ready to take power. One contingent suggests that it's when it'll accelerate without stumbling; another one says it's when the oil temp reaches the bottom of the green. Personally, I have two criteria: the cylinder head temperature has to be well into the green, and stabilized; and the oil temperature has to be at least in the lower yellow.

Takeoff in a Turbocharger

When turbocharged engines are cold, they call for the utmost finesse on the throttle during power advancement for takeoff. Remember, the turbo control system depends on oil for operation, and when the oil is sluggish, there's an excellent chance of sluggish wastegate operation and subsequent engine overboost. If your engines are cold, I suggest the following technique: Take your time with your runup, checking everything that needs checking and generally get your ducks in a row. A few extra moments, at, say, 1500 rpm will hurt the engine a lot less than even a momentary overboost might. When cleared for takeoff, taxi into position, get on the brakes, and slowly bring the power up to around eight in. Hg. below your normal takeoff power. Now stop throttle movement while you release the brakes and get underway. By the time you look back at the manifold pressure, you'll most likely find that the engines have picked up a good five inches, and when you add the remaining three inches, there'll be much less of a tendency to overshoot. The engines may pick up an extra inch or so due to increased ram recovery, too.

If air and engine temperatures are normal, not quite as much care is required. I usually apply power and release the brakes at the same time, with only a momentary hesitation 5 in. Hg. or so short of takeoff power.

Climb

There's no need to be in a great rush to get that first power reduction right after takeoff. Sure, there's a lot of sound and fury going on out there, and the engines sound like they're working hard, and they are, but they are also working within their limits. Even those mills having a 5-minute takeoff power limitation are very conservatively rated. Wait until you're at least 500 feet to make that first reduction, since that's one of the most common points for ane engine failure. Bring the manifold pressures back to where you want them, and then *slowly* adjust the props. Rapid changes in rpm are one of the major causes of crankshaft counterweight failure. Besides that, it scares the passengers. A really smooth executive pilot should be able to make all but the largest power changes without the passengers being aware of them.

Whether to climb at or near best-rate speed or to make a cruise climb depends on a number of factors, including winds and weather. In either case, modern turbocharged engines reduce pilot workload to a minimum, since manifold pressure and fuel mixture generally remain constant without further attention. Despite efforts of the foreign oil producing countries, fuel is still cheaper than cylinders and valves, so I generally run conservatively rich at high climb power settings. This often improves aircraft performance more than just the amount contributed by the higher power itself, since the richer mixture also provides enough additional cooling to allow at least partial closure of the cowl flaps. I'm not suggesting you should run full rich, trailing black smoke like an old 707, but a good 100 to 150 deg. F on the rich side of peak won't cost you much extra fuel.

As altitudes increase, you may find engine temperatures starting to creep up again. This is not necessarily due to the engines' having labored so long to get you up there, but rather because the thin air at high altitudes doesn't do quite as good a job of carrying heat away from the engines. On the other hand, it also produces less parasite drag, so opening the cowl flaps a bit won't harm your performance nearly as much as it would at lower altitudes.

Cruise

Your choice of cruise altitude and power setting will be covered at length later on this book, but even for a given power setting, you are presented with a certain number of options. Generally, I lean toward the most *oversquare* combination allowed by the engine power-setting charts. The slower you run the engines, the quieter everything gets. There's also a school of thought that suggests that slower engine speeds allow the fuel and air mixture more time in the cylinders for more complete and efficient combustion.

Whether or not this latter point is a valid one, I don't know. What I can state with certainty is that even significantly oversquare settings won't damage the engines as long as the manufacturers' limits are honored. Service life may actually be increased, too. Again, opinions differ, but one school of thought feels that every engine has a service life determined basically by "how many turns" it has built into it. Obviously, a turn at idle power will scrape less metal off the bearings than one at a high manifold pressure, but the fact remains that the engine does a lot turning if it's running at a lower rpm.

The cruise regime, of course, is also where mixture control becomes most important. Most manufacturers modify their cruise performance figures to reflect several mixtures, including *best power* (usually a bit on the rich side of peak) and *best economy* (at peak, or even a bit beyond.) Feelings are divided about operation on the lean side of peak EGT, and of course some engines simply won't run smoothly once past the stoichiometric mixture.

One problem you may encounter when leaning near or beyond peak is that the peak itself is higher than the maximum permissible TIT. Those manufacturers that authorize leanside operation provide special leaning procedures. I'll quote the one for the Piper Navajo Chieftain:

"At 260 BHP or less, the engines may be operated at peak EGT or on the lean side of peak EGT, as long as stable engine operation results without exceeding any engine limitations during steady-state or transient conditions.

Instructions for leaning:

- At the desired power setting, lean mixture slowly until EGT has stabilized at peak. *Caution:* Do not exceed 1650 deg. F EGT.

● If peak EGT is 1650 deg. F or less, continue to lean if desired until a maximum of 50 deg. reduction is obtained. Readjust manifold pressure as necessary to maintain the desired power setting.

● If 1650 deg. F is reached before peak EGT is obtained, lean according to the following procedure:

With mixture leaned to 1650 deg. F, reduce manifold pressure until EGT is reduced approximately 75 deg.

Lean mixture slowly until peak EGT is obtained.

Lean mixture additionally until 50 to 100 deg. F on the lean side of peak is obtained. Do not lean into engine roughness.

Slowly increase manifold pressure to desired setting without permitting EGT to exceed 1650 deg. F. Carefully adjust mixture until EGT is 1625 to 1650 deg. F. *Note:* Enriching themixture will*increase* the EGT.

Before enrichening the mixture, reduce the manifold presure as in step (a) to prevent exceeding 1650 deg. F EGT."

For accurate cruise-power settings, you'll have to take temperature variations from standard into account. (This is particularly important for turbine aircraft, by the way.) Moreover, once you've set power, keep an eye on the OAT gauge. It may be necessary to readjust power if you fly into a warmer or colder air mass.

Descent

Descents from cruise to approach or maneuvering altitude are often an area in which the best-laid plans are ruined by ATC. This is unfortunate, since an efficient descent can make a significant contribution to the efficienty of your flight as a whole. Moreover, too-rapid of a descent is one of the most common major engine abuses.

A high-performance recip engine operates at some rather high temperatures, even at cruise. Exhaust gases, and temperatures within the cylinders, can approach or exceed 2000 deg. F. The engine is designed to handle these temperatures, and it does. What shortens an engine's life isn't high temperatures, as such,but rapid *changes* in temperature.

Just consider, for example, the relationship between cylinders and cylinder heads. The jugs themselves are made of steel, but the pistons are aluminum; so are the heads. Different alloys are used for each of them, and one characteristic of different metals is that they expand and contract with temperature at different rates. Have you ever seen how a thermostat works? It uses a *bimetallic strip* of two metals bonded together. Since they expand at different rates, heating the strip causes it to bend in the direction of the strip with a slower expansion-contraction rate, while cooling it makes it bend the other way.

Of course, the metals in a thermostat are chosen to maximize this effect, and while those in an engine are also chosen to minimize it, it's still there. Rapid changes in engine temperature increases the difference in temperature between adjacent components, disallowing them sufficient time to warm or cool one another to the same temperature. Chopping the throttle on a hot engine on a cold day can literally curl the heads up like inchworms, leaving you with goodies like cracked valve guides, sticking valves, egg-shaped pistons, and broken rings.

How do we avoid this? By planning our descents carefully to avoid big power reductions, and what reductions we do make, we'll make in gradual stages, with careful attention to engine temperatures.

The most efficient descent is one made at or near cruise power, with an attentive increase in airspeed. Of course, this is seldom realized, except under VFR conditions; more often, ATC will keep; us higher, longer and then it will want us to descend faster.

When the time comes to start down, then, let's make the smallest possible power reduction, and keep those engines warm. In normally-aspirate ships, you have to enrich the mixture when you start down. In a turbo airplane, you may well have to lean the mixture even further than at cruise, and a number of manufacturers specify a minimum, as well as a maximum, EGT. If cowl flaps are open, close them by all means.

As you start down , airspeed will increase. As long as it doesn't get out of hand, that's fine. We'll just arrive a bit sooner. You may have to make progressive power reduction, then wait until temperatures stabilize before reducing power further.

There are other things that will aid descent, and the higher gear and flap speeds of newer aircraft are a big help here. True, it may look a bit odd to see an airplane descending through 14,000 feet with its feet out, but who's looking) Unless you're flying an Aero Commander, your passengers can't see the wheels anyway! Approach flaps don't produce quite as much drag as wheels do, but they often have higher allowable speeds. They can also tend to make the airplane a bit more stable at lower speeds.

Landing, Taxi, Shutdown

Once in the approach and landing phase, engine management sometimes becomes secondary to the business of flying the airplane. This is as it should be; however, a little attention before the giving gets hot and heavy can spare engines some of the mistreatment they normally are subjected to. You should have a pretty good idea of what the maximum power required on the approach will be, and you can set an appropriate mixture for that before reaching the initial fix. Again, this will tend to keep the engines warm during the approach. Crossing the final fix, it's probably a good idea to have mixtures set for climb power in the event of a miss or go-around.

That final "chop, drop, and stop" maneuver isn't the greatest thing in the world for engines, not to mention the attitude of your passengers. Remember that smoothness is the mark of the professional pilot—*ease* the power off as you cross the fence. Remember those poor little counterweights, sliding back and forth on the crankshaft journals as you change rpm!

Once on the ground, some pilots bring the mixtures back a bit from full rich, ostenibly to prevent "loading up" the engines at idle power. While this may be necessary for some engines, a well-adjusted injection system shouldn't be overly rich at idle anyway, and we've now reached a point at which we may not want to keep the engines all that warm anymore! Instead, let's taxi with them full rich, so that they run as cool as possible before we subject them to the really major cooling where we shut them down! (However, leaning just a bit when taxiing out for takeoff can help the engines warm up, reducing the shock when takeoff power is applied. Just make damn sure they're set properly at full rich before you take the runway!)

5 Flying A Typical Cabin Twin: Piper Navajo

Just to give you an idea what it's like to fly a typical turbocharged cabin twin, let's go for a check flight in one of the most common airplanes in the class—a Piper PA-31 Navajo.

The airplane marked a turning point in the history of Piper Aircraft, since it was the first time that one of their production aircraft had ventured into the area of airplanes larger than the venerable Aztec. The Navajo was an immediate success, and has continued in production in various models, including the basic aircraft, or Navajo C. There is also a version using the same airframe, but with slightly larger counterrotating engines, known as the Navajo CR. There is a "stretched" airplane, the Navajo Chieftain, which has a pair of counter-rotating 350-hp Lycomings. The original non-turbocharged version of the airplane is no longer in production; nor is the pressurized version, the Navajo P, which used geared 425-hp engines, although the same airframe forms the basis of the turboprop Cheyenne.

If anything, the Navajo C is the basic workhorse of the line, and it, together with the Chieftain, which offers two more seats, are the two of the essential crafts to many commuter and part 135 operations. Various interior options are available for the C, ranging from posh executive seating, complete with bar and potty through high-density commuter chairs, to a vast empty space suitable for cargo.

A couple of airframe options can also be ordered to tailor the airplane for specific missions. Normally, baggage is carried in the nose (loaded through a large door on the left side of the nosecone) and in an area aft of the main cabin door. If large amounts of baggage are anticipated, extended engine nacelles are available as an option (standard on the CR and Chieftain) which contain a large baggage locker on each side. If cargo use is intended for the airplane, a couple of extra doors are available. One is a cargo door aft of, and contiguous with, the main cabin door. This not only simplifies access to the aftmost baggage area, but it provides an opening some four feet across (when both cargo and main doors are open) for the loading of bulky cargo. The other door, while not intended for cargo, simplifies operations when the cabin is full of boxes, or mailbags. It's simply a pilot door, an openable version of the large left-hand front window. Access to it is up and over the trailing edge of the left wing, and a suitable area of non-skid surface is installed when the pilot door is fitted.

The Navajo is a very straightforward airplane (Fig. 5-1), with no exotic aerodynamic or systems features. It can be an ideal introduction to the world of heavier twins, since it's very similar in most aspects to lighter aircraft, only it's bigger. Many entry-level pilot positions in the world of commuter and corporate aviation are found in Navajos and similar airplanes of the class, such as the Cessna 402 or older 401 and 411 series.

Navajo Preflight

A systematic preflight can be started at the nose baggage compartment, since this is a handy place to keep a fuel sampler, a couple of rags, and a flashlight; *not* the same one we'll use in flight, of course, First, though, let's open the cabin door (it's a clamshell type, with steps in the lower half), head up to the cockpit, and make a couple of preliminary checks. The master switch is at the base of the circuit breaker panel on the left sidewall. We'll switch it on, and observe the indications of the two fuel gauges in the overhead panel. These gauges only show the quantity of the tanks that are active on the fuel selector, so after we've noted the main tank indications, we'll position the fuel selectors, on the forward face of the main spar, which crosses the floor just aft of the pilot seats, to the aux tanks to verify their fuel quantities as well. Now, let's position the preselect-type flap control to "down," allowing the electrically operated flaps to extend for a look at their tracks, hinges, and pushrods. Finally, let's turn on the two pitot heat switches in the overhead for just a moment; now, turn them off, turn off the master, and leave the cabin.

Assuming we haven't dallied too long in getting back to the front, or pointy, end of the airplane, the two pitot tubes under the nose should still be quite warm to the touch. Let's get our fuel drainer and rag out of the nose baggage compartment and we'll be on our way.

Since we're already at the baggage compartment, let's take a quick look at the hydraulic system fluid reservoir gauge at the back of the compartment; we should be able to see red fluid in the sight gauge. Insufficient fluid can be particularly embarassing in the Navajo; although it's assumed that the emergency standpipe would retain enough fluid for the hand pump. If it, too, were dry, you'd have no way of pulling the hydraulic uplocks for emergency gear extension.

Incidentally, the aircraft battery is located just forward of the baggage compartment, behind a removable panel. If radar is installed, its R/T unit and antenna share the available space in the nose cone. Other avionics are usually installed on a shelf above and at the rear of the nose baggage compartment.

Fig. 5-1. The Piper Navajo is more docile than most corporate twins in it's class. It's often the first airplane you'll fly as a professional pilot (courtesy of Piper).

Now, let's duck under the aircraft and have a look into the nose wheel well, checking for obstructions, damage, and loose parts or connections. We can also check the nose gear itself for proper tire and strut inflation and the condition of all moveable parts and fasteners.

At the root of the left wing, we'll find the first of a number of fuel drain points: a large one for the left main tank, a smaller one for the fuel system filter (this one behind a small access door), and another small one for the pressure crossfeed line. Drain all three, checking for contaminations or water. Some aircraft are equipped with Nayak nacelle tanks (which transfer electrically into the mains.) If they have these tanks, they too, have a drain on the bottom of the inboard side of the nacelle.

Now, the left main gear and wheel well come in for inspection. Again, we'll check the usual items: security and condition of connections and parts, proper inflation of tire and oleo strut. Normally, the Navajo's inner landing gear doors are closed when the gear is fully extended, but it's a good idea to check the condition of the inner door actuators and wheel wells at least once a week. To do this, merely actuate the cockpit emergency hand pump with the master switch off; with no power applied to the gear-door solenoid valve, the hydraulic pressure will cycle them to the open position. (To close them, either pump them back up with the master switch on or just leave them alone; they'll come up when hydraulic pressure comes on after engine start.)

Moving on to the left engine nacelle, we'll check a few items indigenous to Navajos that appear on both sides. The Navajo cowling is one of the easiest to remove, since it simply comes off in a top and bottom half; as a result, the fasteners that keep it from doing so merit some special attention. At the firewall, both upper and lower cowls are attached with small screw fasteners resembling beefy Phillips screws. In my experience, given the forgetfulness of most mechanics, a single such fastener has an average half-life of only a few weeks, so carry a few spares. The rearmost fasteners on the lower cowl half are covered by a little fiberglass wing-leading-edge fillet. Make sure that this is in place and secured by its three screws. Along each side of cowling are three large locks that look like immense slot-head screws. Make sure that these are turned fully clockwise, with their slots horizontal, and that you can see a small pin protruding into the center of each slot. The pin is a lock which allows the "screw" to turn only when a screwdriver is in the slot. Many operators paint a stripe across both the screw and the cowling adjacent to it to provide a positive "locked" indication.

At the frint of the cowling, four more of the small screw fasteners keep top and bottom halves together. This is also your chance to peer into the cowling, checking for possible bird's nests or other obstructions, and to give the alternator V-belt a gentle tug to test for correct tension. Losing the services of an alternator due to a broken belt isn't just an inflight annoyance; changing a belt necessitates removing the propeller! Keeping the belts at proper tension not only assures adequate alternator output, but it also makes the belts last longer.

The big cowl flap is part of the lower half of the removable cowling. Its two actuating rods are attracted to the flap by clevises, clevis pins, and two safety-pin-like Fokker fasteners. Make sure that this entire arrangement is secure, particularly if the airplane has been worked on (cowlings removed) since the last flight.

A pair of small louvered doors on the upper aft cowling offers a means for visual and tactile inspection of the turbocharger and engine accessory section. Bitter experience has taught the author that these doors, if left unsecured, will flap and flutter in the airstream after takeoff until they tear off their hinges, so make sure they're latched properly!

Checking the oil shows us another area in which the Navajo is a bit unusual—the oil dipstick is on the left side of both engines, and since they're canted by the airplane's dihedral angle there are different scales on it for either engine. Both dipsticks have both scales, so make sure you (or line personnel) are looking at the correct one for the correct engine.

Outboard of the nacelle, the flush filler caps for both main and aux tanks adorn the upper wing surface, while the aux tank sump drain is on the lower surface just outboard of the nacelle. Check the de-ice boots for scrapes, nicks, or tears. You might notice the little, sharp-edged rubber stall strips glued to the de-ice boots about a third of the way out to the tip. These have a habit of peeling off, and can be glued back down with neoprene cement. Insignificant as they may appear, they contribute considerably to the airplane's pre-stall warning buffet, and you shouldn't fly without them. The stall-warning vane is on the left leading edge. An ice-inspection light is on the outboard side of the left nacelle.

The Navajo ailerons have mass-balance counterweights at their outboard ends, which move in and out of the fiberglass wingtip fairings through cutouts in their lower surface. Move the aileron through its full travel on preflight to make sure there's no accumulation of crud on the counterweights that could cause aileron binding. Work your way inboard along the trailing edge, checking aileron and flap hinges. If you have wing lockers, make sure they're properly loaded and secured.

The static ports are on the side of the tailcone, just aft of the door, check them, then move out along the stabilizer, noting the condition of de-ice boots, elevator hinges, and trim tab pushrods. Note the small sections of the de-ice boot on the elevator horns (the sections of the elevator, near the outboard ends, that protrude forward of the hinge line.) These prevent ice from forming and jamming in the slot between the elevator horn and the fixed portion of the stabilizer.

The right-hand side of the aircraft is a virtual mirror image of the left. There are a couple of minor differences, including the absence of a crossfeed-line drain on this side, and the presence (in some aircraft) of the optional air-conditioning system, with its electrically-operated airscoop door, in the right nacelle. Remember to check the dipstick scales! Also, when checking the oil, you're in a good position for a quick look at the outside of the aircraft emergency exit door, built around the second large window on the right side of the fuselage. It should be flush with the fuselage; if it isn't be sure to make a double-check of its locking mechanism when you re-enter the cabin. With the preflight complete, stow your supplies in the nose compartment, close and *lock* its door, and climb aboard.

Unlike most other clamshell doors, the Navajo's incorporates its lock mechanism in the bottom half, so after boarding and verifying that the aft cargo door (if installed) is properly fastened, release the top half's hold-open latch and gently allow it to close. Then haul the bottom half up by its cables, and close it, making sure that the cables hang freely in the cabin and aren't pinched in the door. If it's properly closed, the over-center latching mechanism should require almost no force to actuate.

The Navajo Cockpit

Let's settle into the cockpit and look around (Fig. 5-2). There's an overhead panel containing most of the electrical switches and lighting controls, with magnetos, starters, fuel boost pump switches, and fuel and alternator gauges in a "eyebrow" over the windshield. Flight instruments are in front of each pilot, with radios and engine instruments in the center panel (Fig. 5-3). The throttle console also has the autopilot programmer, three-axis trim controls, and cowl flap switches and indicators. On standard-length Navajos, the fuel selector controls are on the front of the main spar as it passes through the cockpit just aft of the seats. Since the Chieftain cockpit is a few feet further forward, its fuel controls are on the floor at the bottom of the throttle console. Don't put anything on the floor between the seats, since that blocks access to the little hatch covering the emergency hydraulic hand pump. Circuit breakers, alternator output switch breakers, and the split master switch are located in the left sidewall. On recent Navajos, all warning lights are neatly consolidated in an annunicator array at the top of the instrument panel. In most others,

Fig. 5-2. Piper Navajo interior (courtesy of Piper).

they're scattered about a bit, with door-open warnings on the eyebrow panel, fuel system warnings at the top of the main instrument panel, and alternator-fail lights virtually out of sight at the bottom of the sidewall switch panel.

There are a few items to be checked before starting (Fig. 5-4), after turning on the master, check that both main and baggage door warning lights are out, and that both alternator failure lights are lit. The two alternator switch breakers should be "on." Position the fuel selectors to "main" for both tanks, make sure the crossfeed handle is "closed" (aft), and verify proper fuel leven indications on the gauges.

It's very important to be sure the landing gear handle is in the "down" or "down neutral" position, since, as soon as an engine is started, hydraulic pressure will come up. If the handle is raised on the ground, the system has sufficient muscle that the gear *will retract*. Piper provides a safeguard in the form of a solenoid lock that physically prevents the handle from moving beyond the "up neutral" position as long as there's weight on the wheels. You can check this by trying to raise the handle with engines stopped: if it moves beyond "up neutral," the system should be checked before flight. Incidentally, if you move the handle into "up neutral" on the ground, the gear warning horn will sound until the handle is placed in the "down neutral" or "down" positions.

Firing Up the Engines

Since the Navajo's battery is more or less equidistant from the engines, there's no particular reason to start one or the other first. Select either engine, advance the throttle about half an inch, set the mixture at full rich, and run the fuel boost pump for a few seconds, or until the fuel *flow* (not pressure) needle stirs from its peg. Older Navajos have only the two standard electric boost pumps; newer ones have two boost pumps in the wings (considered "boost" pumps) and two more, considered "emergency boost pumps," in the engine system plumbing. The wing pumps are normally on whenever there's power on the electrical system, and their failure, which doesn't affect normal flight operations except in case of vapor lock at high altitude, is indicated by appropriate warning lights.

With boost (or, as the case may be, emergency boost) pumps off after priming, return the mixture to idle cutoff, switch on both magnetos for the selected engine, and activate the starter switch (between the mag switches.) As the engine begins to fire, move the mixture control to full rich and watch the oil pressure gauge, which should start indicating rising pressure within 30 seconds.

Now, before starting the second engine, you can check at least one of your two hydraulic pumps. Move the gear handle from "down neutral" to "down." If the inner gear doors were left open after a preflight, you may hear them close now. After a few seconds, the handle should snap back to

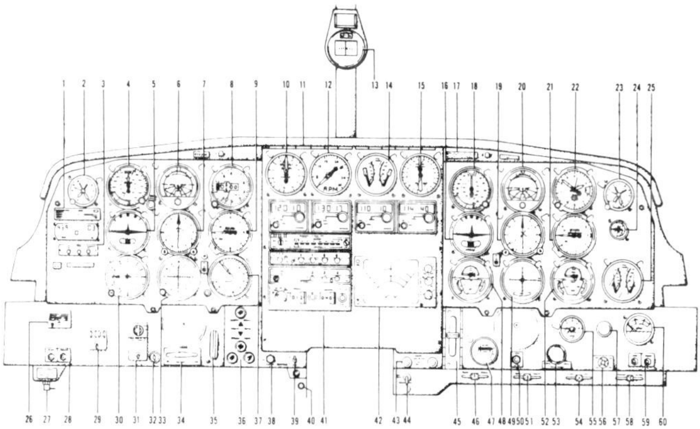

1. AIRSPEED PLACARD
2. CLOCK
3. DME
4. AIRSPEED INDICATOR
5. TURN AND BANK INDICATOR
6. ATTITUDE GYRO
7. DIRECTIONAL GYRO
8. ALTIMETER
9. RATE OF CLIMB INDICATOR
10. DUAL MANIFOLD PRESSURE GAUGE
11. ANNUNCIATOR PANEL
12. DUAL TACHOMETER
13. COMPASS
14. DUAL EGT GAUGE
15. DUAL FUEL FLOW GAUGE
16. TURN AND BANK INDICATOR · COPILOT
17. CATEGORY PLACARD
18. AIRSPEED INDICATOR · COPILOT
19. DIRECTIONAL GYRO · COPILOT
20. ATTITUDE GYRO · COPILOT
21. RATE OF CLIMB INDICATOR · COPILOT
22. ALTIMETER · COPILOT
23. CLOCK
24. GYRO PRESSURE GAUGE
25. DUAL FUEL PRESSURE GAUGE
26. EMERGENCY LOCATOR TRANSMITTER
27. ALTERNATE STATIC SOURCE VALVE
28. MIKE AND EARPHONE JACKS · PILOT
29. AUTOPILOT CONTROLS
30. ADF
31. WINDSHIELD WIPER CONTROL/ EMERGENCY BATTERY PACK
32. CIGAR LIGHTER
33. NAV 2
34. PARKING BRAKE
35. GEAR SELECTOR
36. GEAR INDICATOR LIGHTS
37. RADAR ALTIMETER
38. RADIO LIGHT SWITCH
39. AVIONICS MASTER SWITCH
40. PA VOLUME CONTROL/ON-OFF KNOB
41. AVIONICS
42. RADAR
43. MANUAL ALTERNATE AIR CONTROLS
44. HEATER CONTROL SWITCH
45. WING FLAP SELECTOR SWITCH
46. CABIN TEMP. CONTROL
47. WING FLAP POSITION INDICATOR
48. OIL PRESS., CYL. HEAD TEMP., AND OIL TEMP. · LEFT ENGINE
49. NAV 2 · COPILOT
50. FLAP TEST SWITCH
51. DEFROSTER CONTROL
52. OIL PRESS., CYL. HEAD TEMP., AND OIL TEMP. · RIGHT ENGINE
53. CABIN EXHAUST CONTROL KNOB
54. CABIN AIR CONTROL LEVER
55. OXYGEN SUPPLY PRESSURE GAUGE
56. PANEL LIGHTS SWITCH
57. OXYGEN CONTROL KNOB
58. COCKPIT AIR CONTROL LEVER
59. MIKE AND EARPHONE JACKS COPILOT
60. VOLTMETER

Fig. 5-3. Here's the instrument panel of a late-model Piper Navajo CR.

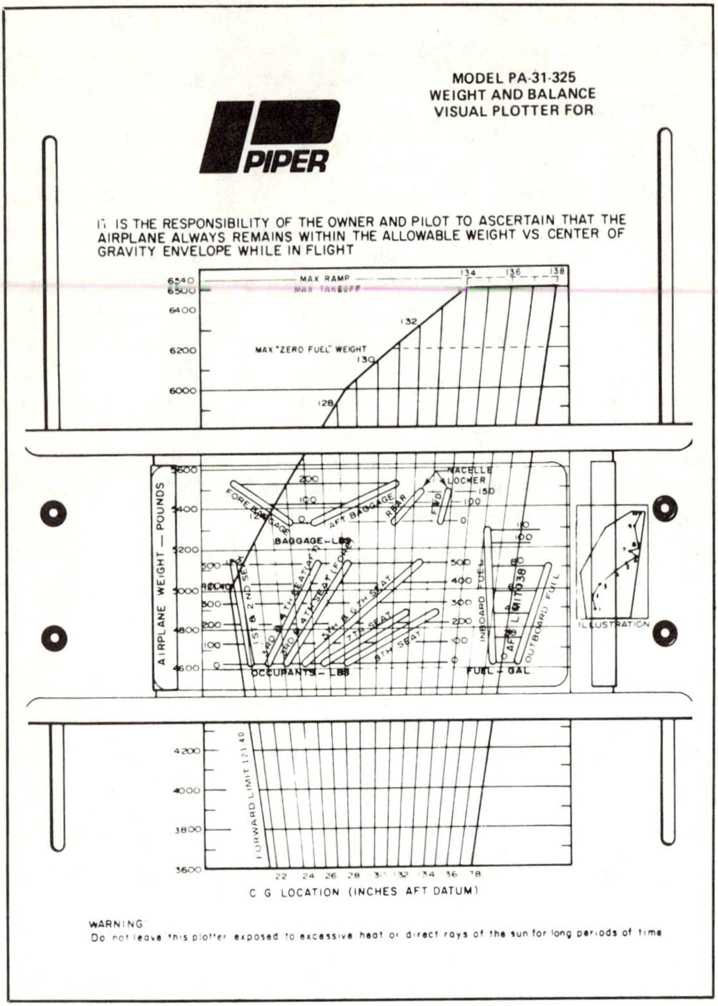

Fig. 5-4. This weight and balance plotter makes determination of exact gross weight and center of gravity a simple process that can be carried out before each flight. Piper provides these with new Navajos; with a little ingenuity, you can make one for any aircraft you fly.

the neutral position by itself, indicating that pressure has risen in the gear system. A very rough check can be made of the second hydraulic pump after starting its engine by again putting the handle down and noting whether it takes less time to snap back than on a single engine. But a more conclusive way of doing it is to remember which engine was started first, then shut that one down first at the end of the flight and check pump operation on the other engine.

Pre-Taxi Steps

Final pre-taxi steps include turning on the avionics, most Navajos have a radio master switch for this, and, if passengers are aboard, verify that all are in their seats and obeying the "no smoking" and "fasten seat belts" signs you may have turned on before starting the engines.

You may be surprised at how ponderously but inexorably a larger airplane keeps on rolling once you get it underway, so I make a habit of discretely checking the brakes before I begin taxiing. This doesn't mean that you have to plaster the passengers against the cabin divider; just inch the throttles up until the airplane barely begins to move, then gently apply the brakes to stop it again. The first turn out of your tiedown spot is a good place to make sure that both turn and bank gyros and the directional gyros, HSI, and/or RMI are working properly. In warm weather, you may have turned on the air conditioning by now; in cold, you may have started the heater. In either case, take a look at the overhead ammeter to make sure that you're not overtaxing your generating capacity at low engine speeds. While you're at it, note the total ammeter reading. Now, one at a time, press the *alternator test* buttons next to the ammeter and verify that each is putting out about the same amount of current. If they're very different, you should have the paralleling system and voltage regulators checked.

Remember Your CIGAR

There are, of course, printed checklists for every phase of operation, and if such are available, follow them by all means. For those cases in which no checklist is immediately available, or when you're switching frequently from one aircraft type to another, there are a few universal mnemonics you can use. My favorite for the before-takeoff check is "CIGAR:" Controls, Instruments, Gas, Attitude, Runup. Let's take them in turn:

Controls. All controls should be free and correct over their entire range, and the flaps should be set for takeoff. Many pilots, when checking the controls, simply move the yoke from one extreme or roll to the other, then they pull it back all the way and allow it to return forward. This procedure only provides real proof that two conditions exist: aileron freedom with the yoke all the way forward, and elevator freedom with the ailerons neutral. A far better technique is to roll the yoke all the way to one side, then hold it there while moving it aft; then roll it over to the other side, and move it forward. Now repeat the procedure, with the initial aileron input the other way, and you can be sure that the controls are , indeed free throughout their range. Full rudder travel is a bit hard to check on airplanes with a lot of weight on the nosewheel, but try anyway.

When it comes to setting the flaps for takeoff. I use the opportunity to check their full travel as well. I extend them all the way, then bring them back up to the takeoff position. If I'm going to have a limit switch fail or an asymmetric flap condition, I want to find out about it *now*, not inflight!

Instruments. In addition to checking the usual gyro heading and altimeter setting, we'll crosscheck against the copilot instruments. If there's a VOT signal or VOR checkpoint at the airport, we'll check the VOR's as well; if there's an error that's still within the allowable legal limits, I've gotten into the habit of noting the correction with a grease pencil right on the panel next to the instrument face. Now let's set the VOR's for departure. I (cIgar) also stands for *Icing*; if icing conditions are expected upon or immediately after takeoff, now is a good time to turn on the heated pitots, prop de-ice and heat the windshield (*after* setting the DG, since the windshield instruments cause large errors in the magnetic compass!)

Gas. We want to be sure, before takeoff, that we can feed from any tank and that the engine-driven pumps work. First, then, let's make sure the electric boost pumps are off. Now, we'll turn the left fuel selector to "off" and monitor fuel pressure. When its starts to drop, open the crossfeed, and it should rise again, with both engines now running from the right main tank. After a few seconds, switch the *right* fuel selector to the aux tank and note that both engines run on the right aux.

Next, switch the *left* fuel selector to "aux," and the right to "off," and monitor continued operation; both engines are now feeding from the left aux tank. After a couple of moments, switch the left fuel selector back to "main," verify continued operation, the close the crossfeed. When right fuel pressure starts to drop, switch the right fuel selector back to main. Check that both firewall shutoff valves are open with their protective covers installed. This all sounds quite complex, but it takes less than 30 seconds. Finally, switch on whatever boost pumps are called for on takeoff, and note the rise in fuel pressure.

Attitude. This term doesn't just mean "are we cheerful?"—it refers here to the correct trim setting. Remember, there are three trim tabs: elevator, rudder, and aileron. A touch of right rudder makes for a smoother takeoff in non-CR Navajos. If there's electric trim, first make sure it works in the right direction, then make sure that its cutoff switch does, indeed, stop it instantly (with trim switch held in either direction).

Runup. This procedure is accomplished in the Navajo in three stages. First, run both engines up to 1500 rpm (runup can be done one engine at a time, but both at once is quicker, smoother, and puts less side load on the nose gear from unequal power).

At 1500 rpm, take a quick look around the panel, making sure all engine instruments are in the green and that vacuum is at the correct level, with neither of the two red *source failed* buttons in view. Now, smoothly and quickly biring both propeller levers around the detend and into the "feather" position. There's no need to wait for any major drop in rpm; at this low engine speed, *any* drop is an indication that the props are feathering, and letting the props get too far into feather will cause all sorts of vibrations, and could even stall an engine. As the props feather and come back up to speed, note whether they both cycle at about the same speed.

Now, slowly work the throttle up to 2300 rpm. This sounds like a high figure, but there are reasons for this. This is a bit past the point at which the turbochargers start to pressurize the system above field ambient, so it's a good check for them. If the oil is cold, you may "overshoot" the desired rpm by a bit, at it'll take a couple of seconds of careful throttle matching to sync the engines up by ear. All right, now they're synchronized. Do the tachs read the same? How about the manifold pressures, fuel flows, and EGT's; anything more than a needle width difference in any parameter suggests that either the instrument or the airplane's fuel injection system and power lever rigging should be checked. Next, carry out the magneto check. Doing both engines at once makes it easy to detect discrepancies by the change in *beat note* as the engines go out of sync.

Taking Off In The Navajo

Cleared for takeoff? We'll line up and take a last quick look at the *runway items*: fuel pumps on, strobes on, transponder on, fllaps set. The throttles can be advanced fairly rapidly to 25 in. Hg, or so, but movement beyond that should be cautious, particularly when the oil is cool. Maximum manifold pressure will be somewhere between 38 and 43 in Hg depending on temperature, since the turbo system's density controller will compensate for higher-than normal temperatures with an increase in manifold pressure. With a V_{mc} of 90 mph, 95 makes a good initial rotation speed, and the airplane will fly off very nicely at a bit over 100. If you're going to go on the dials, 5 degrees nose-up on the gyro horizon is a good target to shoot for while you get the rest of your ducks in a row.

As soon as you're committed to flying, move the gear handle to "up." If it doesn't move past the neutral position, take just an instant to check the circuit breaker panel; if the gear control breaker has popped, the safety solenoid lock on the handle will still be in place. There are a number of big-airliner style thumps and clunks from under the cabin floor to impress the passengers during the gear retraction cycle; you'll know it's complete when the red *gear in transit* light goes out, and in a second or two, the gear handle will pop down to the *up neutral* position.

By now, and with no further action on your part, assuming you've held fairly close to your 5 degree deck angle, the airspeed should be well above V_{yse}, and you can make the transition to a cruise climb configuration. Retracting the flaps will produce a perceptible settling sensation, unless you simultaneously bring the nose up to 8-10 degrees on the attitude gyro. It takes a couple of takeoffs to get it really smooth, but when it's done correctly, only those passengers who've been looking out the window at the wing will know that the flaps have come up. Passing 500 feet above takeoff, you can set 35 in. Hg. and 2400 rpm as a good basic climb power. At normal weights and around 140 mph indicated, this should still give you around 1000 fpm. Manifold pressure will require little or no attention up to about 14,000 feet.

Some moderate leaning can be accomplished as soon as climb power has been set, and here's one regime in which fuel flow takes precedence over EGT as a leaning guide. The manufacturer has established definite minimum fuel flows for the 85% power, and the extra fuel is needed for cooling. It does its job, too. On all but the hottest days, you can close the cowl flaps about halfway during the climb, and the cylinder head and oil temperatures will stay comfortably in the green.

One final task is engaging the propeller synchronizer on those airplanes so equipped. Synch up the engines as closely as possible; for some reason (probably lost motion in the cables) the system seems to work best if the final adjustment is made in the *decrease rpm* direction. When the engines are in sync, turn on the synchrophaser. You may hear a momentary change in engine tone as the unit adjusts the props to the factory-preset phase angle for minimum noise. Power changes can be made with prop sync on if it's done very slowly and carefully. If the synchronizer reaches the end of its limited range, it must be reset by turning it off, resynchronizing, and re-engaging it. It works best if throttle quadrant friction is set fairly tight to prevent "creeping" prop control levers.

Cruising In The Navajo

Once at cruise altitude, we've arrived at the easy part of the job; there's not too much to do but select and set an accurate power figure, switch tanks as required, and monitor aircraft performance and the navigational situation. Since this is a pilot report, however, a bit of airwork might be in order.

The Navajo handles much like smaller twins. Steep turns take a good deal of back pressure, and serve to remind you that you have something bigger than a light trainer strapped to you, but there's no fundamental difference in technique. Stalls are relatively docile, and are presaged by both a warning horn and an unmistakable buffet, especially with the airplane clean. With gear and flaps down, the break is a bit mushier, but the inboard stall strips still ensure that aileron control is available right through the maneuver. An engine cut below V_{mc} during stall practice results in a roll-off to the side of the dead engine, but the big vertical tail makes it a very lazy one. In other words, this is not an airplane that will sneak up and bite you when your back is turned. Otherwise, there's not too much to say about the cruise regime. The cabin is bright and airy, with those big windows, and at the usual oversquare cruise power settings, it's fairly quiet. The cockpit is a little louder, since it's closer to the props, but it's still a very pleasant "office." Forward visibility is excellent; to the sides, you're aware of the big engine nacelles, however, it's still possible for the conscientious pilot to clear traffic. Handling is generally smooth and stable; the airplane's an ideal first step into the plesures of larger equipment.

Landing The Navajo

Letdowns are best started quite a way out; the Navajo is clean, and picks up airspeed quite readily on a descent. Mixtures should be leaned as necessary to keep the EGT above 1350 deg. F during the descent. One aid to steep descents is the forst 15 deg. of flap, which can be lowered at indicated airspeeds of up to 175 mph. With the relatively low indicated speeds at cruise altitude, it's easy to lower the flaps right at the start of a descent, if necessary. There's a moderate pitch change with initial flap extension particularly at higher airspeeds. The rest of the flaps, and the gear, have less effect on the trim.

Before entering, the landing pattern is a good time for preliminary landing checks: radios set properly, fuel on the main tanks and pumps on and passengers secured. While there's no reason a Navajo can't mix with smaller, slower traffic, at larger fields it's very easy to keep speeds up for the convenience of following airline traffic. One hundred and fifty mph is max gear-extension speed. The wheels will come down with a perceptible clunk, particularly the nosewheel. Most Navajos have a small convex mirror mounted on the inboard side of the left engine nacelle to allow the pilot a visual check of nose gear position. At night, the landing lights offer similar confirmation, since they're mounted on the gear leg. (At times this is unfortunate, since it's impossible to "show a light" for traffic avoidance until after the gear is down.)

Pattern speeds are equivalent to the aircraft's general docility; 120 to 140 mph downwind depending on traffic, slowing on base to 110 mph or so, works very well. I like to have all power speed, and configuration changes completed on base leg whenever possible to aid in setting up a stabilized approach as far out on final as possible; at normal weight, an over-the-fence speed of 100 mph works very well. Even at that speed, there's enough kinetic energy for an entirely adequate flare, with power "wiped off" either during or just before the flare. The softest touchdowns seem to be made on the mains with the nose gear about 6 inches off the runway, and at forward CG locations, this can take a fairly decisive back pressure. I sometimes use the electric trim during the flare, although purists may frown on this technique. You'll get a softer nosewheel touchdown if you lower it gently to the runway after landing. If you hold it off as long as possible by main force, you'll "run out of tail" and it'll drop with a bit of a thump.

When you were flying smaller ships, you may have been used to retracting the flaps after touchdown for better braking. In a larger airplane, this usually isn't a good practice. With their higher stall speeds, most larger ships are "all done flying" shortly after touchdown, and they'll have enough weight on the wheels for effective breaking even with the flaps extended. Moreover, in an airplane like the Navajo, with a hydraulic system powerful enough to fold up the casters even with the airplane's weight on them, seeing hands roaming around the gear and/or flap controls after touchdown makes me, for one, very nervous. Leave the flaps alone until after you've turned off the active; then *look* and *feel* to make sure you have the right handle, look *again*, then bring up the flaps. Other after-touchdown chores in the Navajo include turning off the fuel pumps, squawking "standby," and turning off wingtip strobes. If you anticipate shutting down within the next couple of minutes, turn the heater to its "fan" position (if it was used inflight) helps it cool down and prolongs its life.

Abnormal Situations

What about abnormal situations? Again, the Navajo represents a good transition airplane between the light twins and the really "big iron," since it performs and handles in a similar fashion.

Engine failures at cruise are almost anticlimactic; at normal cruise speeds, that big tail provides enough stability to make even the initial, uncorrected swerve fairly minor, and single-engine ceiling performance is adequate.

Engine loss on takeoff is a slightly different matter; and it's compounded by the fact that it's seldom practiced. Most Navajo operators would be just as pleased if their engines never had to face the nasty thermal shock of being chopped after just a few moments of full-power operation. It is sufficient to say that the initial swerve is strong, but no unbearably sudden, and there's enough rudder to keep first things first for as long as it takes to assess the situation. There's just enough rudder trim to cope with a failed engine, and carrying the dead side 5 degrees higher will help a great deal.

The flap preselect mechanism in recent Navajos helps a lot when dealing with an engine failure during a go-around. You can zip the selector up to the 15 degree position, and then you can turn your attention to other things. The flaps will dump most of their drag, but not their lift. I'd still consider the airplane committed to land if an engine failed below 300 feet.

The Navajo is typical of (if not somewhat more docile than) the general membership of the *cabin twin* class. Other members include such machines as the Cessna 401, 411, and 402; the 560, 680 and the "Grand" series Aero Commanders; and the Beach Twin Bonanza and Queen Air. It's a class in which you may well find yourself spending a lot of flying time as an entry-level professional pilot, and you may encounter equipment that's not the newest and most sophisticated around, not to mention passengers ranging from sullen commuters to even more sullen and inert sacks of mail in the still watches of the night. Don't look down on these airplanes, though, the majority of general aviation's "working" miles are probably flown in machines of the ilk, and they're the ideal place in which to sharpen your flying skills to the point that they're so automatic that you can later devote more attention to the systems and management of more sophisticated, higher performing, pressurized aircraft and/or turboprops.

6

Flight Planning For Efficiency

The days are gone, unfortunately, when a pilot could just load up the airplane with passengers and full fuel, make a quick weather check, fire up, and go. Today, between fuel costs and the increasingly complex ATC environment, the pilot of a high-performance twin should become used to a more complete type of flight planning. It'll be more than worth the time and effort in terms of time and fuel saved, not to metion that peace of mind that comes from knowing "you've got all the answers." Moreover, with a little practice, and given the really excellent performance manuals available today, complete and intelligent flight planning needn't add more than 15 minutes to your normal flight-planning process.

A number of factors have to be considered in planning each flight. Sometimes, of course, some particular operational consideration may outweigh all others. For example, if you're flying a corporate airplane and the Chairman of the Board says, "Get me there as fast as you can," you may have to forget about your flight planning. After all, he's paying you! On the other hand, for most operations, a little planning will allow you to economize, at only a slight sacrifice in block-to-block speeds.

General Considerations

Many flight planning considerations will vary from aircraft to aircraft, but there are a few generalizations we can make. We can also add a few pointers toward evolving a set of efficient guidelines for whatever aircraft you'll be flying.

One general rule which may sound paradoxical, applies to missions that bring the airplane near, or beyond, its maximum normal operating range. The rule is: to get there faster, fly slower!

What we mean here is that the airspeed for maximum range is always considerably lower than that of maximum power cruise. For example, for the Cessna 421 Golden Eagle (Figs. 6-1, 6-2 and 6-3), 73.5% power at FL 250 yields a TAS of 241 knots, for an overall range (including start, taxi, takeoff, climb, letdown, and 45 minutes reserve at cruise power) of 882 nm

Fig. 6-1. The Cessna 421 Golden Eagle has a range of 1,483 nm at 192 kts at 25,000 ft. with 1,572 lbs. of fuel. Single-engine service ceiling is 19,850 ft. (courtesy of Cessna).

(1236 lbs useable fuel). Bring the power back to the max-range figure, though, and while TAS drops to 191 knots, range goes out to 1088 nm, an increase of 206 nm.

True, it would take 4:31 to fly 1088 nm at 241 KTAS, and 1:10 longer at the lower airspeed, but the required letdown, fuel stop, and second takeoff and climb necessitated by the higher cruise speed would more than eat up the time difference. There's also the minor matter of a significant difference in fuel consumption. With the added periods of operation at lower, less efficient altitudes, which would be required if a fuel stop were necessary, you might easily burn an additional 30 gallons (currently at around $1.40/gal.—perhaps future readers will look back on these days with a sigh of regret!).

This, then, is one area where you'll have to develop your own guidelines, based on your particular mission profiles. One of the handiest ways to do this is to construct a speed vs. range graph. I personally feel that one hour is the lowest realistic time to consider for a refueling stop, which is an FBO's claim of a "fast turnaround" notwithstanding. Between the time on the ground and the subsequent climb, you can figure on adding an hour to your total time if you have to land enroute for any purpose.

Here's another generalization: for best economy, slow down in tailwinds, and speed up in headwinds and crosswinds.

Why? Well, the longer you spend flying with a tailwind, the more chance it has to help you along. Obviously, the stronger the tailwind, the slower you can fly for the same groundspeed. By the same toke, the less time you spend in a headwind *or crosswind,* the less time it has to perform its evil intent. Why crosswinds as well as headwinds? Because, since you have to crab into a crosswind to maintain the desired ground track, it'll have some effect toward lessening your groundspeed, even if it's from directly abeam. Again, you can construct simple graphs for a quick determination of how much to slow down or speed up; or you can simply flight-plan by groundspeed, then adjust your airspeed accordingly once you're underway, using navigational checks or DME readings for groundspeed.

Beware of FT Reports

A point that might be made here is one that will probably displease some of my (former) friends at the National Weather Service: sometimes the Winds Aloft forecast (FT report) is worth about as much as the paper it's printed on. These reports are generated by a computer model based on observations several hours old, and while the weather folks do their best, there are often considerable discrepancies in their forecasts. In fact, I hereby give Lert's Law of Winds Aloft Forecast Interpretation to you, to pass on to posterity or ignore as you see fit: "Forecast winds aloft are generally accurate with regard to direction. With regard to speed, they're relatively accurate as long as they're crosswinds. Headwinds become proportionally stronger than the forecast indicated the closer they come to being right on your nose, while tailwinds become progressively weaker than forecasted the closer they come to being right on your tail."

Given the uncertainty of FT's, I usually don't spend too much time working out highly accurate cruise airspeed corrections to take advantage of winds unless I'm really trying to stretch the range of the airplane. Instead, I usually use a single rule of thumb: in a tailwind situation, I *reduce* my cruise TAS by half the tailwind component, while in a headwind I *increase* my cruise TAS by at least half the headwind component (more if the airplane can do it. I expect that many pilots, myself included, often simply forget the whole thing (at least on flights were max range isn't a consideration) and just jam the power levels up to their usual 65% in calm air or tailwinds, and maybe 75% in headwinds, and just leave it at that! Frankly, on the average 250-300 nm stage, it won't make more than a few minutes difference anyway.

Choosing a Cruise Altitude

A factor that *can* make a great deal of difference, on the other hand, is the choice of cruise altitude. Again, drawing a couple of simple graphs can

Fig. 6-2. Typical Golden Eagle instrument panel (courtesy of Cessna).

make the correct choice a lot easier. Of course, we're dealing with two separate situations here. One addresses the difference between climb and cruise airspeeds, to optimize the altitude for a flight of given length, while the other considers the effects of winds at altitude, based once again on the FT, whatever that may be worth. (Another generalization: go high eastward, low westward, but remember the old adage: "Climate is what we expect; weather is what we get.")

An interesting point crops up here; again, it's one that varies to some extent from airplane to airplane. It's the difference between a climb at maximum climb power and the more common cruise-climb at higher airspeeds.

In a turbocharged aircraft (or in many turboprops), the higher you go (within limits), the more efficient the aircraft is overall. This suggests that the sooner you reach cruise altitude, the better off you are, and performance figures bear this out. For example, at max climb power, the Golden Eagle requires almost 10 minutes to get to FL 180 from sea level on a standard day. When it levels off, it'll have burned about 80 lbs. of fuel and it will have flown 21 nm along its course.

On the other hand, at cruise-climb power and a higher airspeed, it'll take 18 minutes to get to altitude, burning about 87 lbs of fuel and getting 20 nm further on—41½ nm. For the moment, it looks as if a cruise climb is a better idea, but wait: once leveled off, the ship will be knocking off a hair better than 43 nm per minute. So if two identical Golden Eagles were to launch side by side, with one cruise-climbing and one scrambling for altitude at max climb power, the max-climb airplane would arrive at a point in space 41 nm down the road at FL 210 more than 2 minutes ahead of the cruise-climbing airplane. Fuel burn at cruise is about 3½ lbs/min, so the two airplanes would arrive at the same point in space with a difference in total fuel burn of only a pound or so.

Which method is better, then? Either one, depending, once again, on the particular situation. For example, if you anticipate a tailwind, get on up there as soon as you can. Remember, winds get stronger the higher you go (usually). By the same token, in headwinds, a cruise climb is probably a better idea. Your airspeed will be higher (as we discussed above), and you'll spend a bit more time at lower altitudes where the headwind isn't quite so bad. Onceagain, other considerations may supervene, for example, the Chairman of the Board may not like spending ten minutes with the nose of the airplane pitched way up and the engines howling away outside, although, if proper operational limits are observed,a max power climb shouldn't hurt them.

What about picking the right altitude to climb to in the first place? Well, this is an area that can really become a can of worms. Again, about the only hard and fast rule is that the actual optimum altitude is almost always guaranteed to be unavailable due to weather, wind, or ATC conflicts! Let's take a look at another chart from the Cessna Golden Eagle—the *range profile,* which shows cruise airspeed and range vs. altitudes for

Fig. 6-3. Cessna Golden Eagle interior (courtesy of Cessna).

given power settings (if you want to use a different power setting, the chart can be interpolated).

Obviously, the higher you go, the faster (and hence farther) you can go for every gallon you burn. On the other hand, it takes longer to climb to higher altitudes. A climb to high altitude for a short flight would increase the overall time, since you'd be spending more time at lower airspeeds.

The situation is further complicated by winds at altitude, as well as the differences in airspeeds between cruise and max power climbs. You can calculate graphs to optimize all the factors, but it gets pretty involved; that's why many jets and some turboprops subscribe to computerized flight-planning services. If you're good at math and have one of the new programmable calculators, you can write an optimization program that'll do the job for you—taking inputs of mission distances, head or tailwind components at various altitudes, and cruise and climb airspeeds. Again, it'll have to be written for your own particular aircraft. Let's face it, you can go too far in this flight-planning business! As usual, a rule of thumb can suffice for most operations, and it'll have to be one you develop yourself based on your normal operations and the performance of your airplane. For instance, flying the 421, I generally use a no-wind figure of 1000 feet above an initial 10,000 feet MSL for each additional 50 nm above an initial 100 nm. Below 10,000 feet, the airplane just isn't going to be all that efficient in any case, and the same holds true for flights of less than 100 nm. Thus, in the mythical no-wind condition, I'd plan for 12,000 feet MSL for a 200-nm trip, 16,000 MSL for a 400-mile stage, and so forth. This method is crude, but it works for me. Anyway, as I said, your plans will usually be knocked into a cocked hat by some combination of wind, weather, and ATC.

Table 6-1. A Pre-flight Performance Card.

Date: ________ Departure: ________ Destination: ________
Alternate: ________________
Takeoff: Rotation Speed: ________ Gross Wght: ________
Ground Roll: ________ Ft
Accelerate-Stop Distance: ________________
Accelerate-Go Distance: ________________
Best Single Engine R/C: ________ At ________ Kias
Landing Distance: ________ Ft. @ ________ Kias Approach Speed
(in case an immediate return is required)
Cllimb: R/C ________ FPM @ ________ Kias to ________ Ft Msl
Fuel Consumed: ________ Lbs Distance ________ Nm
Time ________ Minutes
Cruise: ______ Kias (______ Ktas) @ ______ in Hg and ______ RPM
Fuel Flow: ________ PPH
Fuel Consumed: ________ Lbs Distance ________ NM
Time: ________ Hrs/Min
Descent: ________ Kias @ ________ in Hg and ________ RPM
Fuel Consumed: ________ Lbs Distance: ________ NM
Time: ________________ Minutes
Landing (Destination): Gross Weight: ________ Lbs
Final Approach Speed: ______ Kias
Runway Length (50-foot obstacle): ________ Ft
Ground Roll: ________ Ft
Climb: R/C ________ FPM @ ________ Kias to ________ Ft msl
Fuel Consumed: ______ Lbs Distance ______ NM Time ______ Minutes
(climb for cruise to alternate)
Cruise: ________ Kias (KTAS) @ ________ in Hg and ________ RPM
Fuel Flow: ________ PPH Fuel Consumed: ________ Lbs
Distance: ________ NM Time: ________ Hrs/Min
(cruise to alternate)
Descent: ________ Kias @ ________ in Hg and ________ RPM
Fuel Consumed: ______ Lbs Distance: ______ NM ______ Time: ______ Min
Holding Allowance @ ________ in Hg and ________ RPM:
________ Lbs Fuel ________ Hrs/Min
Landing (ALTERNATE): Gross Weight: ________ LBS
Final Approach Speed: ________ KIAS
Runway Length (50-foot obstacle): Ft
Ground Roll: ________ Ft.

Read Your Flight Manual

Equally important to flight-planning is knowing how to extract the required information from your aircraft's flight manual. This is a situation which has continued to improve since the introduction of the standardized GAMA (General Aviation Manufacturers' Association) Flight Manual Format. There's a wealth of information here, if you know how to get it, and as I said earler, it contributes immensely to a pilot's peace of mind to know such things as the fact that there's plenty of room to accelerate and stop or accelerate and go from the takeoff runway. It also helps to know that your single-engine climb capability will be sufficient to overfly any obstacle encountered on take off or any terrain enroute, and also that fuel reserves will be sufficient to reach the primary and/or alternate destinations, with sufficient holding allowance, for even the worst case of adverse winds and weather. Remember another old adage: "The superior pilot is the one who uses his superior knowledge to avoid situations that would require the use of his superior skill."

Performance information is arranged either as graphs or in tabular form. Since tables seem to always have a gap between figures right at the point at which we're going to be flying, there are two methods to use them: *approximation,* in which we enter the tables at the next most conservative values, and *interpolation,* in which we calculate the difference between the actual and tabulated values. For most operations, approximation provides sufficient accuracy, and it's safe. The difference between the table and the actual figure will make airplane performance appear worse than will really be the case. If approximation yields a value that looks uncomfortably close to the limits, then you can embark on the somewhat more laborious and time-consuming interpolation process. As long as you stick with approximation, though, the whole process won't take long, and before each flight you can prepare a performance card as shown in Table 6-1.

The Golden Eagle's Flight Manual

As an illustration of how this is done, we'll turn once again to the Cessna 421 Golden Eagle, since its flight manual is one of the best. In addition to providing all aircraft performance parameters, the manual provides an exhaustively complete sample flight-planning process, which is reproduced here by the kind permission of Cessna Aircraft Corporation.

INTRODUCTION TO TABULATED PERFORMANCE

The performance tables are presented in increments of temperature, altitude and any other variables involved. Performance for a given set of conditions can be approximated as follows:

(1) Takeoff, Accelerate Stop, Accelerate Go, Landing - Enter tables at the next higher increment of weight, altitude and temperature.
(2) Cruise - Enter tables at next lower increment of temperature and altitude.

To obtain exact performance values from the tables, it is necessary to interpolate between the increment values. The following is an example of approximation and interpolation, using an excerpt from the Normal Takeoff Distance Chart.

EXAMPLE

Given:		Find:	
Weight	7100 Pounds	Takeoff Speed	____ KIAS
Temperature	(16°C) 61°F	Ground Roll	____ Feet
Pressure Altitude	2400 Feet	Total Distance to Clear 50-Foot Obstacle	____ Feet
Headwind	19 Knots		

Weight Pounds	Takeoff and Climb Speed KIAS	Pressure Altitude Feet	10°C (50°F)		20°C (68°F)	
			Ground Roll - Feet	Total Distance to Clear 50-Ft - Feet	Ground Roll - Feet	Total Distance to Clear 50-Ft - Feet
7450	100	2000 3000	1930 2050	2490 2650	2130 2270	2810 3000
6800	96	2000 3000	1550 1650	1970 2090	1700 1810	2190 2340

Approximation Method

Extract from the chart the next increment of weight, altitude and temperature which is more conservative than the actual conditions [i.e.: 7450 pounds, 3000 feet and 20°C (68°F)].

Takeoff and Climb Speed . . . . 100 KIAS
Ground Roll . . . . 2270 Feet
Total Distance to Clear 50-Foot Obstacle . . . . . . 3000 Feet

Interpolation Method

If the approximation method yields a value larger than can be tolerated, a more exact value should be determined using the interpolation method.

The example weight (7100 pounds) is 6800 pounds plus 300/650 or .46 times the difference between 6800 pounds and 7450 pounds [i.e.: 6800-pound value + .46 (7450-pound value - 6800-pound value)]

The example pressure altitude (2400 feet) is 2000 feet plus 400/1000 or .4 times the difference between 2000 feet and 3000 feet [i.e.: 2000-foot value + .4 (3000-foot value - 2000-foot value)].

The example temperature of 16°C (61°F) is 10°C plus 6/10 or .6 times the difference between 10°C and 20°C [i.e.: 10°C value + .6 (20°C value - 10°C value)].

Interpolating Values for Normal Takeoff Distance:

Takeoff and Climb Speed = 6800-pound value + [.46 (7450-pound value - 6800-pound value)]

= 96 KIAS + [.46 (100 KIAS - 96 KIAS)]

= 96 KIAS + [1.8 KIAS]

Ground Roll (7 interpolations required)

Altitude interpolation at 10°C (50°F) and 7450 pounds = 2000-foot value + [.4 (3000-foot value - 2000-foot value)]

= 1930 feet + [.4 (2050 feet - 1930 feet)]

= 1930 feet + [48 feet]

= 1978 feet

Altitude interpolation at 20°C (68°F) and 7450 pounds = 2000-foot value + [.4 (3000-foot value - 2000-foot value)]

= 2130 feet + [.4 (2270 feet - 2130 feet)]

= 2130 feet + [56 feet]

= <u>2186 feet</u>

Altitude interpolation at 10°C (50°F) and 6800 pounds

= 2000-foot value + [.4 (3000-foot value - 2000-foot value)]

= 1550 feet + [.4 (1650 feet - 1550 feet)]

= 1550 feet + [40 feet]

= <u>1590 feet</u>

Altitude interpolation at 20°C (68°F) and 6800 pounds

= 2000-foot value + [.4 (3000-foot value - 2000-foot value)]

= 1700 feet + [.4 (1810 feet - 1700 feet)]

= 1700 feet + [44 feet]

= <u>1744 feet</u>

The Normal Takeoff Distance chart, with altitude interpolation, looks as follows:

Weight Pounds	Takeoff and Climb Speed KIAS	Pressure Altitude Feet	10°C (50°F)		20°C (68°F)	
			Ground Roll - Feet	Total Distance to Clear 50-Ft - Feet	Ground Roll - Feet	Total Distance to Clear 50-Ft - Feet
7450	100	2400	1978	----	2186	----
6800	96	2400	1590	----	1744	----

Weight interpolation at 10°C (50°F) and 2400 feet

= 6800-pound value + [.46 (7450-pound value - 6800-pound value)]

= 1590 feet + [.46 (1978 feet - 1590 feet)]

= 1590 feet + [178 feet]

= <u>1768 feet</u>

Weight interpolation at 20°C (68°F) and 2400 feet

= 6800-pound value + [.46 (7450-pound value - 6800-pound value)]

= 1744 feet + [.46 (2186 feet - 1744 feet)]

= 1744 feet + [203 feet]

= <u>1947 feet</u>

The Normal Takeoff Distance chart, with altitude and weight interpolation, looks as follows:

Weight Pounds	Takeoff and Climb Speed KIAS	Pressure Altitude Feet	10°C (50°F)		20°C (68°F)	
			Ground Roll - Feet	Total Distance to Clear 50-Ft - Feet	Ground Roll - Feet	Total Distance to Clear 50-Ft - Feet
7100	98	2400	1768	----	1947	----

Temperature interpolation at 2400 feet and 7100 pounds = $10^{\circ}C$ ($50^{\circ}F$) value + [.6 ($20^{\circ}C$ ($68^{\circ}F$) value - $10^{\circ}C$ ($50^{\circ}F$) value)]

= 1768 feet + [.6 (1947 feet - 1768 feet)]

= 1768 feet + [107 feet]

= <u>1875 feet</u>

The Normal Takeoff Distance chart, with altitude, weight and temperature, looks as follows:

Weight Pounds	Takeoff and Climb Speed KIAS	Pressure Altitude Feet	$16^{\circ}C$ ($61^{\circ}F$)	
			Ground Roll - Feet	Total Distance to Clear 50-Ft - Feet
7100	98	2400	1875	----

Ground Roll with 19-knot headwind

$$= 1875 \text{ feet} - [1875 \text{ feet} \left(\frac{19 \text{ knots headwind}}{10 \text{ knots headwind}}\right)(7\%)]$$

= 1875 feet - 249 feet

= <u>1626 feet</u>

Total Distance to Clear 50-Foot Obstacle (7 interpolations required)

The interpolations required are identical to the ground roll interpolations, except "total distance to clear 50-foot obstacle" values are substituted for the "ground roll" values.

The interpolated value for the total distance to clear 50-foot obstacle is 2432 feet (no wind) and 2109 feet (19-knot headwind).

SAMPLE FLIGHT

The following is an example of a typical flight using the performance data contained in Figures 6-4 through 5-18. The approximation method is used in tabular performance except where noted.

AIRPLANE CONFIGURATION

Airplane Weight . . . . 7100 Pounds
Usable Fuel Load . . . . 1236 Pounds

TAKEOFF AIRPORT CONDITIONS

Field Length . . . . 6000 Feet (Runway 23)
Temperature . . . . $16^{\circ}C$ ($61^{\circ}F$)
Field Pressure Altitude . . . . 2400 Feet
Wind . . . . 270^{0} at 25 Knots
Obstacles . . . . None

CRUISE CONDITIONS

Distance . . . . 600 Nautical Miles
Cruise Altitude . . . . 17,500 Feet
Temperature . . . . $-10^{\circ}C$ ($14^{\circ}F$)
Wind . . . . 15-Knot Tailwind
Power . . . . Maximum Recommended Cruise Power at Recommended Lean Mixture

LANDING AIRPORT CONDITIONS

Field Length . . . . 3500 Feet (Runway 19)
Temperature . . . . 7°C (45°F)
Field Pressure Altitude . . . . 1700 Feet
Wind . . . . 210° at 17 Knots
Landing Weight . . . . To be Calculated
Obstacles . . . . 50-Foot Trees

SAMPLE CALCULATIONS

Wind Component Chart (Figure 6-4

(1) The angle between the runway and the prevailing wind is 40°.
(2) Enter Figure 5-9 on the 40° wind line and proceed out to the intersection with the 25-knot arc.
(3) Read horizontally left from this intersection; the headwind component is 19 knots.

Normal Takeoff Distance (Figure 6 5)

(1) Enter Figure 6-5 at 7450 pounds weight; the takeoff and climb speed is 100 KIAS.
(2) Proceed horizontally right from 3000-foot pressure altitude to the vertical columns for 20°C (68°F). The takeoff ground run is 2270 feet and the total distance required to clear a 50-foot obstacle is 3000 feet without wind correction. With a 19-knot headwind component, the corrected takeoff ground run is 1968 feet and the corrected total distance required is 2601 feet.

$$\frac{\text{19 Knots Headwind}}{\text{10 Knots Headwind}} (7\%) = 13.3\%$$

Corrected Takeoff Ground Run = 2270 feet - [13.3% (2270 feet)]

= 2270 feet - [302 feet]

= <u>1968 feet</u>

Corrected Total Distance Required = 3000 feet - [13.3% (3000 feet)]

= 3000 feet - [399 feet]

= <u>2601 feet</u>

Accelerate Stop Distance (Figure 6-6)

(1) Enter Figure 6-6 at 7450 pounds weight; engine failure speed is 100 KIAS.
(2) Proceed horizontally right from 3000-foot pressure altitude to the vertical columns for 20°C (68°F). The distance required to accelerate to 100 KIAS and stop is 4350 feet without wind correction. With a 19-knot headwind component, the accelerate stop distance can be reduced by:

$$\frac{\text{19 Knots Headwind}}{\text{4 Knots Headwind}} (3\%) = 14.25\%$$

Corrected Accelerate Stop Distance = 4350 feet - [14.25% (4350 feet)]

= 4350 feet - [620 feet]

= <u>3730 feet</u>

Accelerate Go Distance (Figure 6-7)

(1) Enter Figure 6-7 at 7450 pounds weight; engine failure speed is 100 KIAS.

(2) Proceed horizontally right from 3000-foot pressure altitude to the vertical columns for 20°C (68°F). The distance required to clear a 50-foot obstacle, after losing an engine at 100 KIAS, is 13,540 feet without wind correction. With a 19-knot headwind component, the distance can be reduced by:

$$\frac{\text{19 Knots Headwind}}{\text{10 Knots Headwind}} \ (6\%) = 11.4\%$$

Corrected Accelerate Go Distance = 13,540 feet - [11.4% (13,540 feet)]

= 13,540 feet - [1544 feet]

= 11,996 feet

> **NOTE**
>
> - The distance required to accelerate go using the approximation method is so great, in view of the 6000-foot runway available, that a more exact value should be obtained using the interpolation method.
> - The interpolation method gives an accelerate go distance of 6025 feet without wind or 5338 feet with 19 knots of headwind.

Rate-Of-Climb — Maximum Climb (Figure 6-8)

(1) Enter Figure 6-8 at 16°C (61°F).
(2) Proceed vertically up to the 2400-foot pressure altitude line.
(3) Proceed horizontally right to the reference line. Follow the slope of the adjacent rate-of-climb lines until intersecting the vertical 7100-pound line.
(4) Proceed horizontally right to obtain rate-of-climb. (1925 Feet per minute)
(5) Enter the climb speed data to determine the climb speed corrected for 7100 pounds and 2400 feet. (108 KIAS)

Rate-Of-Climb — Cruise Climb (Figure 6-9)

(1) Enter Figure 6- 9 at 16°C (61°F).
(2) Proceed vertically up to the 2400-foot pressure altitude line.
(3) Proceed horizontally right to the reference line. Follow the slope of the adjacent rate-of-climb lines until intersecting the vertical 7100-pound line.
(4) Proceed horizontally right to obtain rate-of-climb. (1260 Feet per minute)
(5) Climb speed is 120 KIAS for all conditions.

Rate-Of-Climb — Single Engine (Figure 6-10)

(1) Enter Figure 6-10 at 16°C (61°F).
(2) Proceed vertically up to the 2400-foot pressure altitude line.
(3) Proceed horizontally right to the reference line. Follow the slope of the adjacent rate-of-climb lines until intersecting the vertical 7100-pound line.
(4) Proceed horizontally right to obtain rate-of-climb. (335 Feet Feet per minute)
(5) Enter the climb speed data to determine the climb speed corrected for 7100 pounds and 2400 feet. (109 KIAS)

Time, Fuel And Distance To Climb — Cruise Climb (Figure 6-14)

Time, fuel and distance to climb are determined by finding the difference between the airport and the cruise conditions; thus, two calculations are required, one for the airport condition and the second for the cruise condition.

Airport Condition:
(1) Enter Figure 6-14 at 16°C (61°F).
(2) Proceed vertically up to 2400-foot pressure altitude line.
(3) Proceed horizontally right to the 7100-pound line.
(4) Proceed vertically down to obtain time to climb (2.2 minutes), fuel to climb (11 pounds) and distance to climb (5 nautical miles).

Cruise Condition:
(5) Enter Figure 6-14 at -10°C (14°F).
(6) Proceed vertically up to 17,500-foot pressure altitude line.
(7) Proceed horizontally right to the 7100-pound line.
(8) Proceed vertically down to obtain time to climb (17.4 minutes), fuel to climb (81 pounds) and distance to climb (40 nautical miles).

Final Calculations:

Time to Climb = Cruise time to climb - Airport time to climb

= 17.4 minutes - 2.2 minutes

= <u>15.2 minutes</u>

Fuel to Climb = Cruise fuel to climb - Airport fuel to climb

= 81 pounds - 11 Pounds

= <u>70 pounds</u> (add 46 pounds for start, taxi and runup) (116 pounds total)

Distance to Climb = Cruise distance to climb - Airport distance to climb

= 40 nautical miles - 5 nautical miles

= 35 nautical miles

Adjusted for wind (use 60% of the wind at altitude for climb wind),

= 35 nautical miles ± wind contribution

$$= 35 + \left[\frac{15.2 \text{ minutes}}{60 \text{ minutes}} (.6 \times 15 \text{ knots})\right]$$

= 35 nautical miles + 2.3 nautical miles

= <u>37.3 nautical miles</u>

Time, Fuel And Distance To Descend (Figure 6-17)

Time, fuel and distance to descend are determined by finding the difference between the cruise and the landing airport conditions; thus two calculations are required, one for the cruise condition and the second for the landing airport condition.

Cruise Condition:
(1) Enter Figure 6-17 at the cruise altitude of 17,500 feet.
(2) Proceed horizontally right to the guideline.
(3) Proceed vertically down to obtain time to descend (18.5 minutes), fuel to descend (52 pounds) and distance to descend (55.5 nautical miles).

Landing Airport Condition:

(4) Enter Figure 6-17 at the airport altitude of 1700 feet.
(5) Proceed horizontally right to the guideline.
(6) Proceed vertically down to obtain time to descend (2.1 minutes), fuel to descend (6 pounds) and distance to descend (6.5 nautical miles).

Final Calculations:

Time to Descend = Cruise time to descend - Airport time to descend

= 16.3 minutes - 2.1 minutes

= 14.2 minutes

Fuel to Descend = Cruise fuel to descend - Airport fuel to descend

= 52 pounds - 6 pounds

= 46 pounds

Distance to Descend = Cruise distance to descend - Airport distance to descend.

= 55.5 nautical miles - 6.5 nautical miles

= 49.0 nautical miles

Adjusted for wind (use 40% of the wind at altitude for descent wind),

= 49.0 ± wind contribution

$$= 49.0 + \left[\frac{14.2 \text{ minutes}}{60 \text{ minutes}} (.4 \times 15 \text{ knots})\right]$$

= 49.0 nautical miles + 1.4 nautical miles

= 50.4 nautical miles

Cruise Performance With Recommended Lean Mixture

Maximum recommended cruise can be obtained with 1900 RPM and 32.5 Inches Hg. manifold pressure.

The approximation method for extracting data from the cruise tables is to select the next lower temperature and altitude values, which are generally conservative with respect to fuel economy.

(1) Enter the 15,000-foot data at 1900 RPM and 32.5 Inches Hg. manifold pressure.
(2) Use $-15^{\circ}C$ ($5^{\circ}F$) data for a power of 73.5%, airspeed of 214 KTAS and a total fuel flow of 257 pounds per hour.
(3) Correcting for a weight of 7100 pounds, the airspeed increases to:

$$214 \text{ KTAS} + \frac{(7450 \text{ pounds} - 7100 \text{ pounds})}{1000 \text{ pounds}} (6 \text{ KTAS}) =$$

214 KTAS + 2.1 KTAS = 216 KTAS

Using the interpolation method, interpolating altitude, temperature and weight, the actual performance is 71.2% power, 221 KTAS and total fuel flow of 250 pounds per hour.

In the above calculations, for convenience, the weight was assumed to be equal to the takeoff weight of 7100 pounds. More realistic data can be determined if the average cruise weight is used. This average cruise weight is determined as follows:

$$\text{Cruise Fuel} = \frac{\text{Total distance} - \text{climb distance} - \text{descent distance}}{\text{True airspeed} \pm \text{wind correction}} \times [\text{Total fuel flow per hour}]$$

$$= \frac{600 \text{ Nautical Miles} - 37.3 \text{ Nautical Miles} - 50.4 \text{ Nautical Miles}}{221 \text{ KTAS} + 15 \text{ Knot Tailwind}} \times [250 \text{ pounds per hour}]$$

$$= \frac{512.3 \text{ Nautical miles}}{236} \times 250 \text{ pounds per hour}$$

$= 2.17$ hours x 250 pounds per hour

= <u>543 pounds</u>

$$\text{Average Cruise Weight} = \text{Takeoff weight} - \text{start, taxi and climb fuel} - \frac{\text{Cruise fuel}}{2}$$

$$= 7100 \text{ pounds} - 116 \text{ pounds} - \frac{543 \text{ pounds}}{2}$$

= <u>6713 pounds</u>

Average Cruise Speed = True airspeed from Figure 5-20 + weight correction

$$= 221 \text{ KTAS} + 6 \left(\frac{387}{1000} \right)$$

= <u>223 KTAS</u>

Average Ground Speed = 223 KTAS + tailwind

= 223 KTAS + 15 knots

= <u>238 knots</u>

Distance During Cruise = Total distance - Climb distance - Descent distance

= 600 - 37.3 - 50.4

= <u>512.3 Nautical Miles</u>

$$\text{Time During Cruise} = \frac{\text{Cruise distance}}{\text{ground speed}}$$

$$= \frac{512.3}{238}$$

= <u>2.15 hours</u>

Normal Landing Distance (Figure 8-18)

Landing Weight = Takeoff weight - climb fuel - cruise fuel - descent fuel

= 7100 pounds - 116 pounds - 543 pounds - 46 pounds

= <u>6395 pounds</u>

Wind = 210° at 17 knots. Determine headwind component from Figure 5-9. (16 knots headwind)

Enter Figure 5-25 at 6600 pounds; the approach speed is 96 KIAS. Proceed horizontally right from 2000-foot pressure altitude to the vertical column for 10°C (50°F). The landing distance ground roll is 630 feet and the total distance required to clear a 50-foot obstacle is 2210 feet without wind correction. With a 16-knot headwind component, the corrected ground roll distance is 554 feet and the corrected total distance required is 1945 feet.

$$\frac{16 \text{ Knots Headwind}}{4 \text{ Knots Headwind}} (3\%) = 12\%$$

Corrected Landing Ground Roll = 630 feet - [12% (630)]
= 630 feet - 76 feet
= 554 feet

Corrected Total Distance Required = 2210 - [12% (2210)]
= 2210 feet - 265 feet
= 1945 feet

Rate-Of-Climb — Balked Landing Climb (Figure 6-11)

(1) Enter Figure 6-11 at 7°C (45°F).
(2) Proceed vertically up to the 1700-foot pressure altitude line.
(3) Proceed horizontally right to the weight reference line. Follow the guidelines up and to the right until intersecting the vertical 6395-pound weight line.
(4) Proceed horizontally right to determine the rate-of-climb. (1490 Feet per minute)

Total Fuel Required = Start, taxi and climb fuel + cruise fuel + descent fuel
= 116 pounds + 543 pounds + 46 pounds = 705 pounds (Without Holding Fuel)
or 705 pounds + 125 pounds = 830 pounds (With 45 Minutes Holding Fuel)

Holding Time (Figure 6-16)

To determine holding time, the fuel available for holding must be determined.

Fuel Available for Holding = Initial fuel - [start, taxi and climb fuel + cruise fuel + descent fuel]
= 1236 pounds - [116 pounds + 543 pounds + 46 pounds]
= 531 pounds

(1) Enter Figure 6-16 at 531 pounds of fuel available.
(2) Proceed vertically up to the intersection with the guideline.
(3) Proceed horizontally left to obtain holding time available. (3.2 hours)

WIND COMPONENT

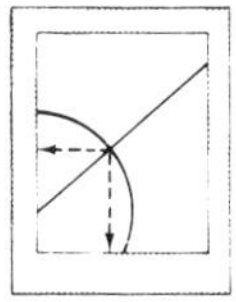

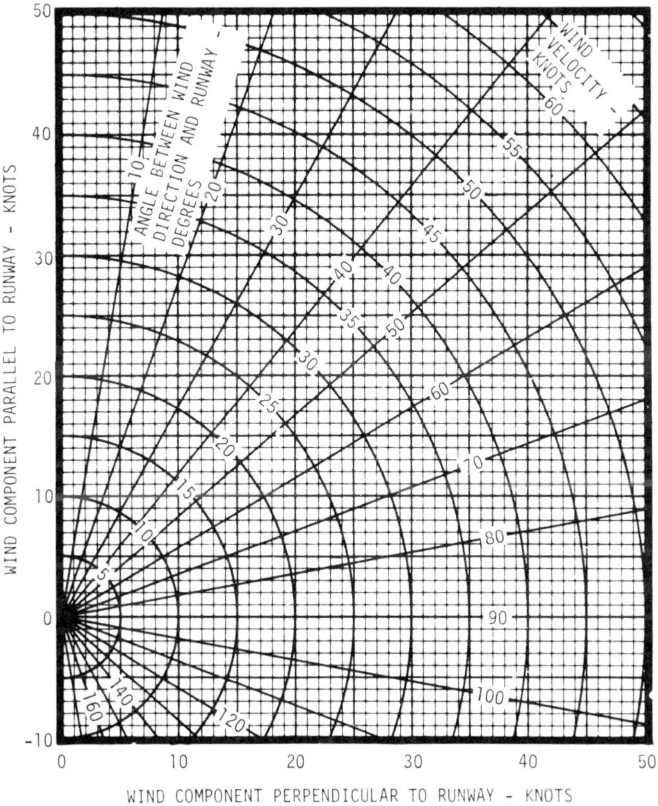

Figure 6-4

NORMAL TAKEOFF DISTANCE

CONDITIONS:
1. 2235 RPM and 39.0 Inches Hg. Manifold Pressure Before Brake Release.
2. Mixtures - CHECK Fuel Flows In the White Arc.
3. Wing Flaps - UP.
4. Level, Hard Surface, Dry Runway.

NOTES:
1. If full power is applied without brakes set, distances apply from point where full power is applied.
2. Decrease distance 7% for each 10 knots headwind.
3. Increase distance 4% for each 2 knots tailwind.

WEIGHT-POUNDS	TAKEOFF TO 50-FOOT OBSTACLE SPEED-KIAS	PRESSURE ALTITUDE-FEET	-20°C (-4°F)		-10°C (14°F)		0°C (32°F)		10°C (50°F)	
			GROUND ROLL - FEET	TOTAL DISTANCE TO CLEAR 50 FEET	GROUND ROLL - FEET	TOTAL DISTANCE TO CLEAR 50 FEET	GROUND ROLL - FEET	TOTAL DISTANCE TO CLEAR 50 FEET	GROUND ROLL - FEET	TOTAL DISTANCE TO CLEAR 50 FEET
7450	100	Sea Level	1280	1610	1410	1780	1550	1980	1710	2200
		1000	1360	1710	1500	1890	1650	2100	1810	2340
		2000	1450	1810	1590	2000	1750	2220	1930	2490
		3000	1540	1920	1690	2120	1860	2360	2050	2650
		4000	1630	2030	1800	2260	1980	2520	2190	2830
		5000	1740	2160	1920	2400	2110	2680	2330	3020
		6000	1850	2300	2040	2560	2250	2860	2490	3230
		7000	1970	2440	2180	2720	2400	3060	2650	3460
		8000	2110	2610	2320	2910	2570	3270	2840	3720
		9000	2250	2780	2480	3110	2740	3510	3030	4010
		10,000	2410	2970	2660	3330	2930	3770	3250	4330
6800	96	Sea Level	1010	1280	1110	1410	1220	1550	1370	1750
		1000	1070	1350	1180	1490	1330	1670	1460	1850
		2000	1140	1430	1280	1610	1410	1770	1550	1970
		3000	1240	1540	1360	1700	1500	1880	1650	2090
		4000	1320	1630	1450	1800	1590	2000	1750	2220
		5000	1400	1730	1540	1910	1690	2120	1870	2360
		6000	1490	1840	1640	2030	1800	2260	1990	2520
		7000	1590	1950	1750	2160	1920	2400	2120	2690
		8000	1700	2080	1870	2300	2050	2560	2270	2880
		9000	1810	2210	1990	2460	2190	2740	2420	3080
		10,000	1930	2360	2130	2620	2350	2930	2590	3300
6200	91	Sea Level	810	1030	890	1130	980	1240	1070	1370
		1000	860	1090	950	1200	1040	1320	1140	1450
		2000	920	1150	1010	1270	1100	1390	1240	1570
		3000	980	1220	1070	1340	1200	1500	1320	1660
		4000	1040	1290	1160	1450	1280	1590	1400	1760
		5000	1130	1390	1240	1530	1360	1690	1490	1870
		6000	1200	1480	1320	1630	1450	1800	1590	1990
		7000	1280	1570	1400	1730	1540	1910	1690	2120
		8000	1360	1670	1500	1840	1640	2030	1810	2260
		9000	1450	1770	1600	1960	1750	2170	1930	2410
		10,000	1550	1890	1700	2080	1870	2310	2060	2570
5600	86	Sea Level	640	820	700	900	770	980	840	1080
		1000	680	860	750	950	820	1040	890	1140
		2000	720	910	790	1000	870	1100	950	1210
		3000	770	970	840	1060	920	1160	1010	1280
		4000	820	1020	900	1120	980	1230	1070	1360
		5000	870	1080	950	1190	1040	1310	1170	1460
		6000	930	1150	1020	1260	1130	1410	1240	1550
		7000	990	1220	1100	1360	1210	1500	1320	1650
		8000	1070	1310	1170	1440	1290	1590	1410	1750
		9000	1140	1400	1250	1530	1370	1690	1510	1870
		10,000	1220	1480	1340	1630	1470	1800	1610	1990

Figure 6-5

NORMAL TAKEOFF DISTANCE

CONDITIONS:
1. 2235 RPM and 39.0 Inches Hg. Manifold Pressure Before Brake Release.
2. Mixtures - CHECK Fuel Flows In the White Arc.
3. Wing Flaps - UP.
4. Level, Hard Surface, Dry Runway.

NOTES:
1. If full power is applied without brakes set, distances apply from point where full power is applied.
2. Decrease distance 7% for each 10 knots headwind.
3. Increase distance 4% for each 2 knots tailwind.

WEIGHT-POUNDS	TAKEOFF TO 50-FOOT OBSTACLE SPEED-KIAS	PRESSURE ALTITUDE-FEET	20°C (68°F)		30°C (86°F)		40°C (104°F)	
			GROUND ROLL - FEET	TOTAL DISTANCE TO CLEAR 50 FEET	GROUND ROLL - FEET	TOTAL DISTANCE TO CLEAR 50 FEET	GROUND ROLL - FEET	TOTAL DISTANCE TO CLEAR 50 FEET
7450	100	Sea Level	1880	2470	2080	2800	2300	3230
		1000	2000	2630	2210	3000	2450	3490
		2000	2130	2810	2360	3210	2620	3780
		3000	2270	3000	2510	3450	2790	4110
		4000	2420	3210	2680	3720	2980	4500
		5000	2580	3450	2860	4030	3190	4970
		6000	2750	3710	3060	4380	3410	5570
		7000	2940	4000	3270	4800	3650	6380
		8000	3150	4330	3500	5290	3910	7570
		9000	3370	4710	3750	5910	4200	9690
		10,000	3610	5160	4020	6730	4510	15,290
6800	96	Sea Level	1510	1940	1660	2180	1840	2460
		1000	1600	2060	1770	2320	1950	2630
		2000	1700	2190	1880	2470	2080	2820
		3000	1810	2340	2000	2640	2220	3020
		4000	1930	2490	2130	2820	2370	3250
		5000	2060	2660	2280	3020	2530	3500
		6000	2190	2840	2430	3240	2700	3790
		7000	2340	3040	2600	3490	2890	4130
		8000	2500	3260	2780	2760	3090	4520
		9000	2680	3500	2970	4080	3310	4990
		10,000	2870	3770	3180	4430	3550	5570
6200	91	Sea Level	1180	1520	1300	1690	1460	1920
		1000	1280	1640	1410	1820	1550	2040
		2000	1360	1740	1500	1930	1650	2170
		3000	1450	1840	1590	2060	1760	2320
		4000	1540	1960	1700	2190	1880	2480
		5000	1640	2080	1810	2340	2000	2650
		6000	1750	2220	1930	2490	2140	2840
		7000	1860	2370	2060	2670	2280	3050
		8000	1990	2530	2200	2860	2440	3290
		9000	2130	2700	2350	3060	2610	3550
		10,000	2270	2890	2510	3300	2790	3850
5600	86	Sea Level	920	1190	1010	1310	1110	1460
		1000	980	1260	1080	1390	1190	1550
		2000	1040	1330	1140	1470	1290	1670
		3000	1110	1410	1240	1590	1370	1780
		4000	1210	1520	1320	1690	1460	1890
		5000	1280	1620	1410	1800	1550	2010
		6000	1360	1720	1500	1910	1660	2150
		7000	1450	1830	1600	2040	1770	2290
		8000	1550	1950	1710	2170	1890	2450
		9000	1650	2070	1820	2320	2010	2630
		10,000	1770	2210	1950	2480	2150	2820

Figure 6-5

ACCELERATE STOP DISTANCE

CONDITIONS:
1. 2235 RPM and 39.0 Inches Hg. Manifold Pressure Before Brake Release.
2. Mixtures - CHECK Fuel Flows In the White Arc.
3. Wing Flaps - UP.
4. Level, Hard Surface, Dry Runway.
5. Engine Failure at Engine Failure Speed.
6. Idle Power and Maximum Effective Braking After Engine Failure.

NOTE:
1. If full power is applied without brakes set, distances apply from point where full power is applied.
2. Decrease distance 3% for each 4 knots headwind.
3. Increase distance 5% for each 2 knots tailwind.

WEIGHT - POUNDS	ENGINE FAILURE SPEED - KIAS	PRESSURE ALTITUDE - FEET	TOTAL DISTANCE - FEET						
			-20°C -4°F	-10°C +14°F	0°C 32°F	+10°C +50°F	+20°C +68°F	+30°C +86°F	+40°C +104°F
7450	100	Sea Level	2900	3090	3290	3510	3750	4010	4300
		1000	3030	3240	3450	3680	3940	4210	4520
		2000	3180	3390	3620	3870	4140	4430	4760
		3000	3340	3560	3800	4060	4350	4670	5020
		4000	3500	3740	4000	4270	4580	4910	5290
		5000	3680	3930	4200	4500	4820	5180	5580
		6000	3860	4130	4420	4740	5080	5470	5900
		7000	4060	4350	4660	4990	5360	5770	6240
		8000	4280	4580	4910	5260	5660	6100	6600
		9000	4510	4830	5180	5560	5980	6450	6990
		10,000	4750	5090	5460	5870	6320	6830	7410
6800	96	Sea Level	2330	2480	2640	2850	3040	3240	3470
		1000	2440	2600	2800	2990	3190	3400	3650
		2000	2560	2760	2940	3130	3350	3580	3840
		3000	2710	2890	3080	3290	3520	3760	4040
		4000	2850	3040	3240	3460	3700	3960	4260
		5000	2990	3190	3400	3640	3890	4170	4490
		6000	3140	3350	3580	3830	4100	4400	4730
		7000	3300	3530	3770	4030	4320	4640	5000
		8000	3470	3710	3970	4250	4560	4900	5290
		9000	3650	3910	4180	4480	4810	5180	5590
		10,000	3850	4120	4420	4730	5090	5480	5920
6200	91	Sea Level	1890	2010	2140	2280	2430	2590	2790
		1000	1980	2110	2240	2390	2570	2750	2930
		2000	2080	2210	2350	2530	2700	2880	3090
		3000	2180	2320	2500	2660	2840	3030	3250
		4000	2280	2460	2620	2800	2980	3190	3420
		5000	2420	2580	2750	2940	3140	3360	3600
		6000	2540	2710	2890	3090	3300	3540	3800
		7000	2670	2850	3050	3250	3480	3730	4010
		8000	2810	3000	3210	3430	3670	3930	4230
		9000	2960	3160	3380	3610	3870	4150	4470
		10,000	3120	3330	3560	3810	4090	4390	4730
5600	86	Sea Level	1500	1600	1700	1800	1920	2040	2180
		1000	1570	1670	1780	1890	2010	2140	2290
		2000	1650	1750	1870	1980	2110	2250	2430
		3000	1730	1840	1960	2080	2220	2390	2560
		4000	1810	1930	2060	2190	2360	2520	2690
		5000	1900	2030	2160	2330	2480	2650	2830
		6000	2000	2130	2290	2440	2610	2790	2980
		7000	2100	2260	2410	2570	2750	2940	3150
		8000	2230	2380	2540	2710	2890	3090	3320
		9000	2350	2500	2670	2850	3050	3260	3500
		10,000	2470	2640	2820	3010	3220	3450	3700

Figure 6-6

ACCELERATE GO DISTANCE

CONDITIONS:
1. 2235 RPM and 39.0 Inches Hg. Manifold Pressure Before Brake Release.
2. Mixtures - CHECK Fuel Flows In the White Arc.
3. Wing Flaps - UP.
4. Level Hard Surface Dry Runway.
5. Engine Failure At Engine Failure Speed.
6. Propeller Feathered and Landing Gear Retracted During Climb.
7. Maintain Engine Failure Speed Until Clear of Obstacle.

NOTE:
1. If full power is applied without brakes set, distances apply from point where full power is applied.
2. Decrease distance 6% for each 10 knots headwind.
3. Increase distance 2% for each 1 knots of tailwind.
4. Distance in boxes represent rates of climb less than 50 ft/min.

WEIGHT - POUNDS	ENGINE FAILURE - SPEED - KIAS	PRESSURE ALTITUDE FEET	TOTAL DISTANCE TO CLEAR 50-FOOT OBSTACLE - FEET						
			-20°C -4°F	-10°C +14°F	0°C 32°F	+10°C +50°F	+20°C +68°F	+30°C +86°F	+40°C +104°F
7450	100	Sea Level	2390	2770	3290	4120	5800	12,210	------
		1000	2550	2980	3590	4630	7020	------	------
		2000	2740	3220	3950	5280	9100	------	------
		3000	2940	3500	4390	6190	13,540	------	------
		4000	3170	3830	4940	7570	------	------	------
		5000	3440	4220	5670	9990	------	------	------
		6000	3750	4710	6710	15,590	------	------	------
		7000	4120	5340	8330	------	------	------	------
		8000	4570	6190	11,350	------	------	------	------
		9000	5130	7430	------	------	------	------	------
		10,000	5870	9480	------	------	------	------	------
6800	96	Sea Level	1770	2000	2270	2670	3180	4010	5770
		1000	1880	2120	2460	2870	3470	4510	7070
		2000	2000	2290	2640	3100	3820	5170	9360
		3000	2150	2450	2830	3370	4240	6080	14,690
		4000	2290	2620	3050	3690	4780	7480	------
		5000	2450	2810	3310	4060	5490	9990	------
		6000	2620	3030	3600	4530	6480	16,070	------
		7000	2810	3270	3950	5120	8040	------	------
		8000	3020	3560	4370	5900	10,930	------	------
		9000	3270	3890	4890	7040	------	------	------
		10,000	3550	4280	5560	8880	------	------	------
6200	91	Sea Level	1380	1530	1710	1930	2200	2570	3120
		1000	1460	1620	1810	2050	2380	2800	3420
		2000	1540	1720	1930	2210	2550	3030	3780
		3000	1630	1820	2080	2360	2740	3290	4220
		4000	1730	1960	2210	2530	2960	3610	4780
		5000	1860	2080	2360	2720	3210	3980	5540
		6000	1970	2220	2520	2920	3500	4450	6650
		7000	2100	2370	2710	3160	3840	5050	8470
		8000	2240	2540	2910	3430	4250	5860	12,200
		9000	2390	2720	3150	3750	4760	7060	------
		10,000	2560	2930	3410	4130	5420	9060	------
5600	86	Sea Level	1070	1180	1300	1440	1610	1820	2090
		1000	1130	1240	1370	1530	1710	1940	2240
		2000	1190	1310	1450	1620	1820	2070	2430
		3000	1260	1390	1540	1720	1930	2240	2610
		4000	1330	1470	1630	1820	2090	2390	2820
		5000	1400	1550	1730	1960	2230	2570	3050
		6000	1490	1650	1860	2090	2380	2760	3330
		7000	1580	1770	1980	2230	2550	2990	3650
		8000	1690	1880	2110	2380	2740	3240	4050
		9000	1800	2000	2250	2550	2960	3540	4540
		10,000	1910	2140	2400	2740	3200	3890	5170

Figure 6-7

RATE-OF-CLIMB - MAXIMUM CLIMB

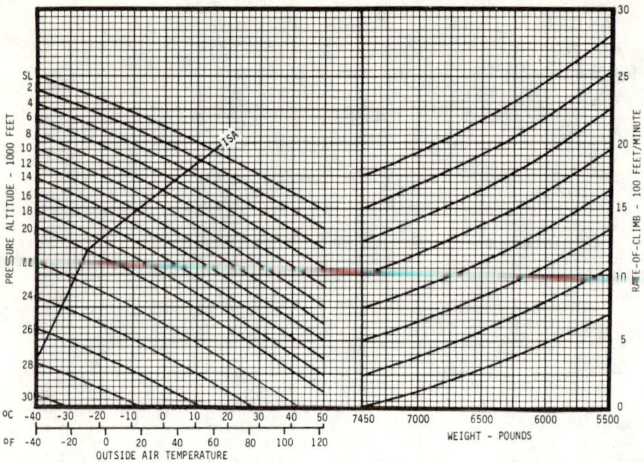

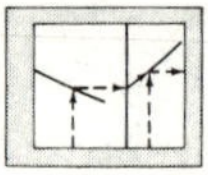

CONDITIONS:
1. 2235 RPM and 39.0 Inches Hg. to 20,000 Feet. Use Placarded Manifold Pressure Above 20,000 Feet.
2. Landing Gear - UP.
3. Wing Flaps - UP.
4. Mixture at Recommended Fuel Flow.

WEIGHT POUNDS	CLIMB SPEED - KIAS		
	Sea Level	20,000 Feet	30,000 Feet
7450	111	105	100
6800	106	99	95
6200	99	95	91
5600	95	90	86

Figure 6-8

RATE-OF-CLIMB - CRUISE CLIMB

CONDITIONS:
1. 1900 RPM and 32.5 Inches Hg.
2. Landing Gear - UP.
3. Wing Flaps - UP.
4. Airspeed - 120 KIAS.
5. Mixtures - Recommended Fuel Flow.

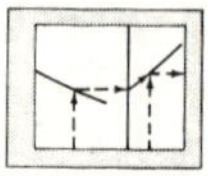

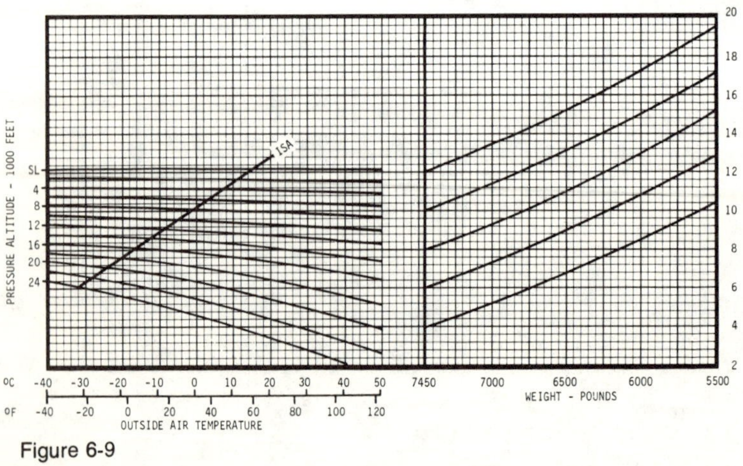

Figure 6-9

RATE-OF-CLIMB - ONE ENGINE INOPERATIVE

WEIGHT POUNDS	CLIMB SPEED - KIAS		
	Sea Level	10,000 Feet	20,000 Feet
7450	111	107	104
6800	109	104	101
6200	106	102	99
5600	105	100	96

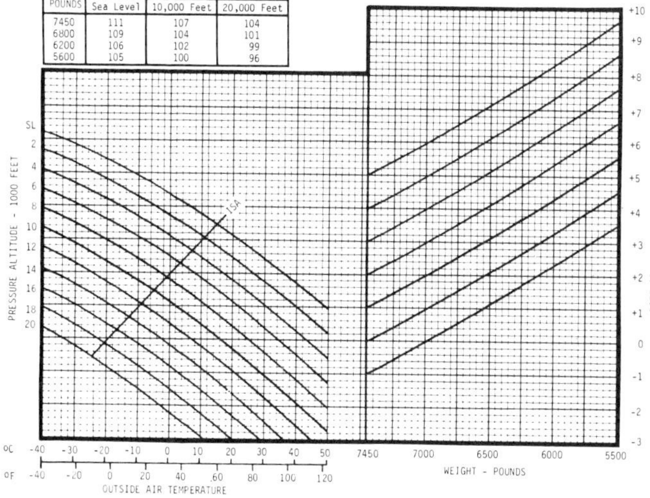

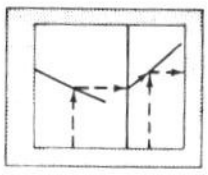

CONDITIONS:
1. 2235 RPM and 39.0 Inches Hg. to 20,000 Feet. Use Placarded Manifold Pressure Above 20,000 Feet.
2. Mixture - CHECK Fuel Flow In the White Arc.
3. Landing Gear - UP.
4. Wing Flaps - UP.
5. Inoperative Propeller - FEATHERED.
6. Wings Banked 5° Toward Operative Engine With Approximately 1/2 Ball Slip Indicated on the Turn and Bank Indicator.

NOTE:
Approximate Effect of Configuration on Single-Engine Rate-of-Climb.

Subtract values listed below from value obtained in above graph. Effects for a combination of gear, flap or windmilling propeller may be obtained by adding the effects for each.

Inoperative Engine Windmilling	400 Ft/Min
Gear Down	350 Ft/Min
Flaps Down 15°	200 Ft/Min
Flaps Down 45°	800 Ft/Min

51847017

Figure 6-10

RATE-OF-CLIMB - BALKED LANDING CLIMB

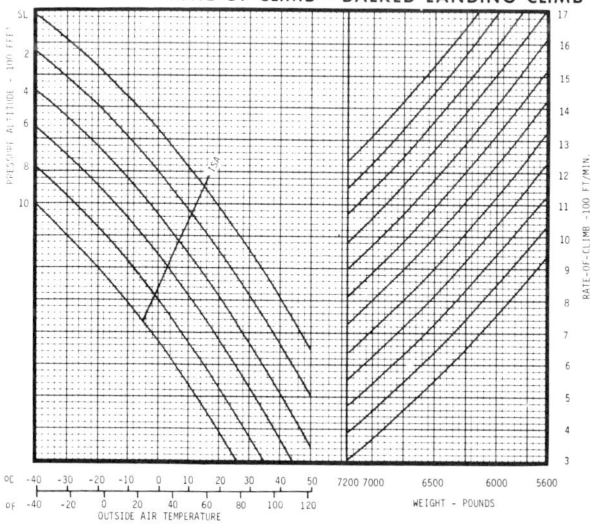

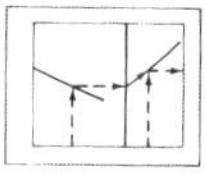

CONDITIONS:
1. 2235 RPM and 39.0 Inches Hg.
2. Mixtures - CHECK Fuel Flows In The White Arc.
3. Landing Gear - DOWN.
4. Wing Flaps - 45°.
5. Climb Speed - 96 KIAS.

Figure 6-11

ENGINE INOPERATIVE SERVICE CEILING

CONDITIONS:
1. Engine Inoperative Climb Configuration.

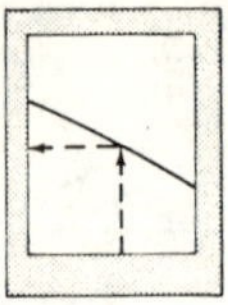

NOTE:
1. Engine inoperative service ceiling is the maximum altitude where the airplane has the capability of climbing 50 feet per minute with one engine inoperative and feathered.
2. Increase indicated service ceiling 100 feet for each 0.10 inches Hg. altimeter setting greater than 29.92.
3. Decrease indicated service ceiling 100 feet for each 0.10 inches Hg. altimeter setting less than 29.92.
4. This chart provides performance information to aid in route selection when operating under FAR 135.181 and 91.119 requirements.

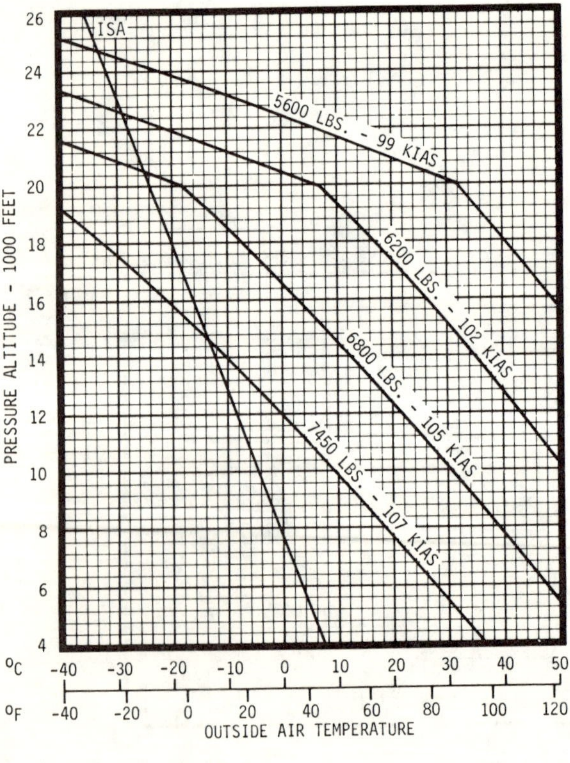

Figure 6-12

TIME, FUEL AND DISTANCE TO CLIMB - MAXIMUM CLIMB

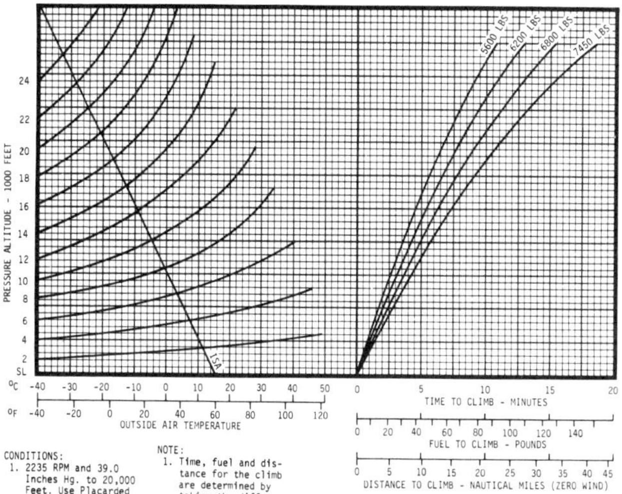

CONDITIONS:
1. 2235 RPM and 39.0 Inches Hg. to 20,000 Feet. Use Placarded Manifold Pressure Above 20,000 Feet.
2. Landing Gear - UP.
3. Wing Flaps - UP.
4. Mixture at Recommended Fuel Flow.

NOTE:
1. Time, fuel and distance for the climb are determined by taking the difference between the airport altitude and initial cruise altitude conditions.
2. For total fuel used, add 46 pounds for start, taxi and takeoff.

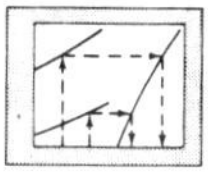

Figure 6-13

TIME, FUEL AND DISTANCE TO CLIMB - CRUISE CLIMB

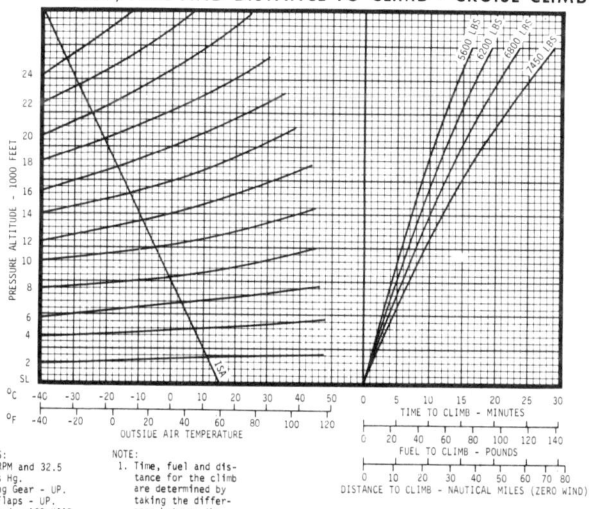

CONDITIONS:
1. 1900 RPM and 32.5 Inches Hg.
2. Landing Gear - UP.
3. Wing Flaps - UP.
4. Airspeed - 120 KIAS.
5. Fuel Flow - BLUE TRIANGLE.

NOTE:
1. Time, fuel and distance for the climb are determined by taking the difference between the airport altitude and initial cruise altitude conditions.
2. For total fuel used, add 46 pounds for start, taxi and takeoff.

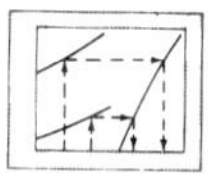

Figure 6-14

ENDURANCE PROFILE

CONDITIONS:
1. Takeoff Weight - 7450 Pounds.
2. Cruise Clumb to Desired Altitude.
3. Recommended Lean Fuel Flow.
4. Standard Day.

NOTE:
1. Endurance computations include fuel required for start, taxi, takeoff, climb, cruise, descent and 45 minutes reserve fuel at the particular cruise power.
2. The endurance shown is the sum of the times to climb, cruise and descend.

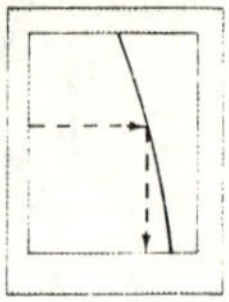

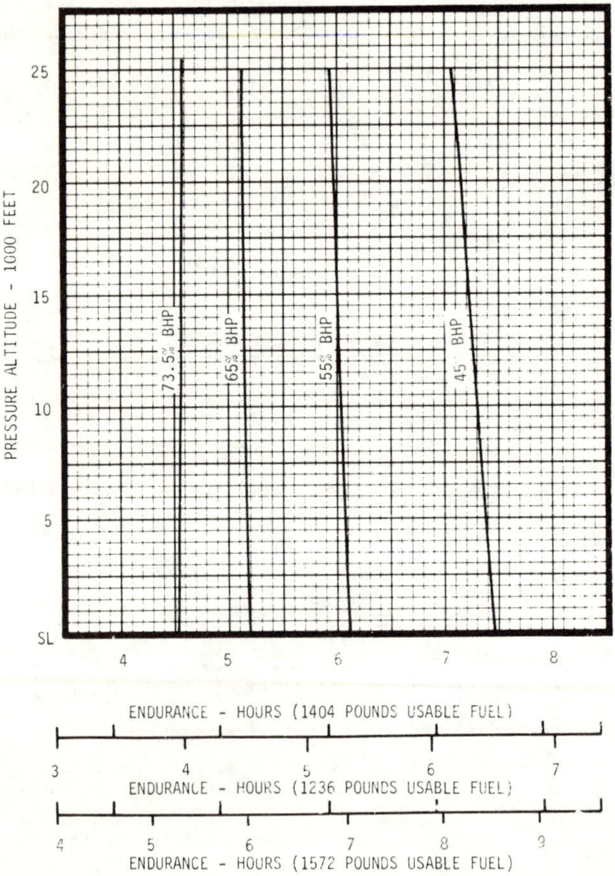

Figure 6-15

HOLDING TIME

CONDITIONS:
1. 1800 RPM and 23 Inches Hg. Manifold Pressure (45% Power).
2. Recommended Lean Fuel Flow (166 Pounds Per Hour Total).

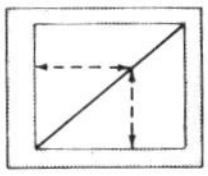

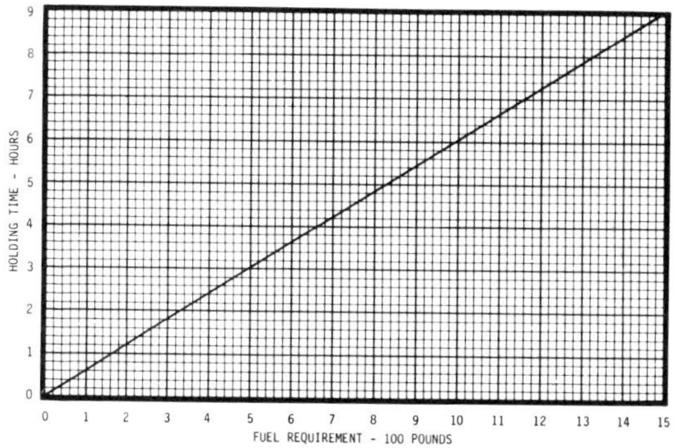

Figure 6-16

TIME, FUEL AND DISTANCE TO DESCEND

CONDITIONS:
1. Power - 1800 RPM and 23 Inches Hg. Manifold Pressure (45% Power).
2. Fuel Flow - RECOMMENDED LEAN (Approximately 83.0 Pounds Per Hour Per Engine).
3. Landing Gear - UP.
4. Wing Flaps - UP.
5. Airspeed - 180 KIAS.

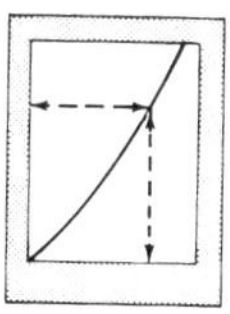

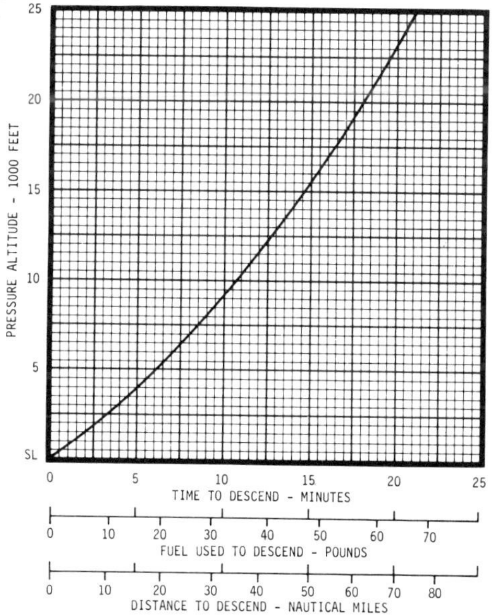

Figure 6-17

NORMAL LANDING DISTANCE

CONDITIONS:
1. Throttles - IDLE at 50 feet above ground level.
2. Landing Gear - DOWN.
3. Wing Flaps - 45°.
4. Touchdown - FULL STALL.
5. Level, Hard Surface Runway.
6. Maximum Effective Braking.

NOTE:
1. If necessary to land with wing flaps UP, the approach speed should be increased above the normal approach speed by 12 knots. Expect total landing distance to increase by 35%.
2. Decrease total distances by 3% for each 4 knots headwind. For operations with tailwinds up to 10 knots, increase total distances by 8% for each 3 knots wind.

WEIGHT-POUNDS	SPEED AT 50-FOOT OBSTACLE KIAS	PRESSURE ALTITUDE - FEET	-20°C (-4°F)		-10°C (14°F)		0°C (32°F)		10°C (50°F)	
			GROUND ROLL - FEET	TOTAL DISTANCE TO CLEAR 50-FOOT OBSTACLE	GROUND ROLL - FEET	TOTAL DISTANCE TO CLEAR 50-FOOT OBSTACLE	GROUND ROLL - FEET	TOTAL DISTANCE TO CLEAR 50-FOOT OBSTACLE	GROUND ROLL - FEET	TOTAL DISTANCE TO CLEAR 50-FOOT OBSTACLE
7200	100	Sea Level	640	2210	660	2240	690	2260	710	2290
		1000	660	2230	690	2260	710	2290	740	2310
		2000	690	2260	710	2290	740	2310	770	2340
		3000	710	2280	740	2310	770	2340	790	2370
		4000	740	2310	770	2340	800	2370	820	2400
		5000	770	2340	800	2370	830	2400	860	2430
		6000	790	2370	830	2400	860	2430	890	2460
		7000	820	2400	860	2430	890	2460	920	2490
		8000	860	2430	890	2460	920	2500	960	2530
		9000	890	2460	920	2500	960	2530	990	2570
		10,000	920	2500	960	2530	1000	2570	1030	2610
6600	96	Sea Level	530	2100	550	2120	570	2140	590	2160
		1000	550	2120	570	2140	590	2160	610	2180
		2000	570	2140	590	2160	610	2180	630	2210
		3000	590	2160	610	2180	630	2210	660	2230
		4000	610	2180	630	2210	660	2230	680	2250
		5000	630	2210	660	2230	680	2260	710	2280
		6000	660	2230	680	2260	710	2280	730	2310
		7000	680	2250	710	2280	740	2310	760	2340
		8000	710	2280	740	2310	760	2340	790	2360
		9000	740	2310	760	2340	790	2370	820	2400
		10,000	760	2340	790	2370	820	2400	850	2430
6000	91	Sea Level	430	2000	450	2020	460	2040	480	2050
		1000	440	2020	460	2030	480	2050	500	2070
		2000	460	2030	480	2050	500	2070	510	2090
		3000	480	2050	500	2070	510	2090	530	2110
		4000	500	2070	510	2090	530	2110	550	2130
		5000	510	2090	530	2110	550	2130	570	2150
		6000	530	2110	550	2130	580	2150	600	2170
		7000	550	2130	580	2150	600	2170	620	2190
		8000	580	2150	600	2170	620	2190	640	2220
		9000	600	2170	620	2190	640	2220	670	2240
		10,000	620	2190	650	2220	670	2240	690	2270
5400	86	Sea Level	340	1910	350	1930	370	1940	380	1950
		1000	350	1930	370	1940	380	1950	390	1970
		2000	370	1940	380	1950	390	1970	410	1980
		3000	380	1950	390	1970	410	1980	420	2000
		4000	390	1970	410	1980	420	2000	440	2010
		5000	410	1980	420	2000	440	2010	460	2030
		6000	420	2000	440	2010	460	2030	470	2050
		7000	440	2010	460	2030	470	2050	490	2060
		8000	460	2030	480	2050	490	2070	510	2080
		9000	470	2050	490	2070	510	2090	530	2100
		10,000	490	2070	510	2090	530	2110	550	2120

Figure 6-18

NORMAL LANDING DISTANCE

CONDITIONS:

1. Throttles - IDLE at 50 feet above ground level.
2. Landing Gear - DOWN.
3. Wing Flaps - 45°.
4. Touchdown - FULL STALL.
5. Level, Hard Surface Runway.
6. Maximum Effective Braking.

NOTE:

1. If necessary to land with wing flaps UP, the approach speed should be increased above the normal approach speed by 12 knots. Expect total landing distance to increase by 35%.
2. Decrease total distances by 3% for each 4 knots headwind. For operations with tailwinds up to 10 knots, increase total distances by 8% for each 3 knots wind.

WEIGHT-POUNDS	SPEED AT 50-FOOT OBSTACLE KIAS	PRESSURE ALTITUDE - FEET	20°C (68°F)		30°C (86°F)		40°C (104°F)	
			GROUND ROLL - FEET	TOTAL DISTANCE TO CLEAR 50-FOOT OBSTACLE	GROUND ROLL - FEET	TOTAL DISTANCE TO CLEAR 50-FOOT OBSTACLE	GROUND ROLL - FEET	TOTAL DISTANCE TO CLEAR 50-FOOT OBSTACLE
7200	100	Sea Level	740	2310	760	2340	790	2360
		1000	760	2340	790	2360	820	2390
		2000	790	2370	820	2390	850	2420
		3000	820	2400	850	2420	880	2450
		4000	850	2430	880	2460	910	2480
		5000	890	2460	920	2490	950	2520
		6000	920	2490	950	2520	980	2550
		7000	950	2530	990	2560	[illegible]	2590
		8000	990	2560	1020	2600	[illegible]	2630
		9000	1030	2600	1060	2640	[illegible]	2670
		10,000	1070	2640	1110	2680	1140	2720
6600	96	Sea Level	610	2180	630	2200	650	2220
		1000	630	2210	650	2230	680	2250
		2000	660	2230	680	2250	700	2270
		3000	680	2250	700	2280	730	2300
		4000	710	2280	730	2300	750	2330
		5000	730	2300	760	2330	780	2350
		6000	760	2330	790	2360	810	2380
		7000	790	2360	820	2390	840	2420
		8000	820	2390	850	2420	870	2450
		9000	850	2420	880	2450	910	2480
		10,000	880	2460	910	2490	940	2520
6000	91	Sea Level	500	2070	510	2090	530	2100
		1000	510	2090	530	2100	550	2120
		2000	530	2110	550	2120	570	2140
		3000	550	2130	570	2140	590	2160
		4000	570	2150	590	2170	610	2180
		5000	590	2170	610	2190	630	2210
		6000	620	2190	640	2210	660	2230
		7000	640	2210	660	2240	680	2260
		8000	670	2240	690	2260	710	2280
		9000	690	2260	710	2290	740	2310
		10,000	720	2290	740	2320	770	2340
5400	86	Sea Level	390	1970	410	1980	420	1990
		1000	410	1980	420	2000	440	2010
		2000	420	2000	440	2010	450	2020
		3000	440	2010	450	2030	470	2040
		4000	460	2030	470	2040	490	2060
		5000	470	2050	490	2060	500	2080
		6000	490	2060	510	2080	520	2100
		7000	510	2080	530	2100	540	2120
		8000	530	2100	550	2120	560	2140
		9000	550	2120	570	2140	590	2160
		10,000	570	2140	590	2160	610	2180

Figure 5-25 (Sheet 2 of 2)

Figure 6-18

7 High Performance And Solo IFR Techniques

If you learned your instrument flying skills in a single (as who didn't) or even in one of the smaller twins, prepare for a pleasant surprise—from here on, it's going to get easier and easier. All that business about having to be a super aviator to handle the IFR environment and ATC relations in a high-performance aircraft is a myth, carefully fostered by the pilots of those very same high-performance aircraft! True, you have to know what you're doing, and a larger aircraft may be no more forgiving of errors than a small one, and true, at higher airspeeds, you cover more ground per unit time, and navigation has to be precise, but, and it's a very big "but", larger aircraft, with their increased wing loading and stability and more sophisticated and reliable systems, make the whole job much easier. Once you realize this, you'll stand in awe of those valiant students and CFII's, slogging through the sky in their 172's and Warriors, bouncing around in turbulence that you barely feel, peering through their hoods at crude, shaky gauges, hoping that their tired old radios don't give up on them and that their aircraft's performance is sufficient to cope with winds and weather. Meanwhile, you're sitting in a great big twin, probably with turbochargers and de-ice, letting the autopilot and sophisticated, steady radios do most of the work while you leisurely monitor the situation, heck, you've got it made!

Even so, there are a few tips that may make your job easier, particularly if you're the only driver up in the pointy end of the airplane, as is the case for most of the recips and a few of the kerosene blowers. Of course, these techniques are good operating practice in any case, and there's no reason why they can't be carried over into the operation of even the lightest single.

Keeping Your Cool

The whole secret behind single-pilot IFR (the factor that lets some pilots step, cool and collected, from the cockpit of even a light single after an approach to minimums, while other pilots may stagger from the flight decks of their big twins and barely make it as far as the airport bar) is simply "staying ahead of the situation" at all times. There's virtually nothing that's that difficult or tiring, as long as you expect it. On the other hand, even something relatively trivial can tumble your personal gyros if it takes you by surprise. The technique, then, is to try to anticipate any surprises and "defuse" them before they ever happen. Then, if something unforeseen does occur, you'll be in the best possible position to cope with it.

We've already dealt with preflight and flight-planning operations in the last chapters. This information can go far toward helping you keep your cool because you already know the following: the condition of your airplane, what the weather is and what it's forecast to be, how much fuel you're going to use, and when you're going to arrive at checkpoints.

Keep a Flight Log

The next step, and it should, of course, still be part of your preflight, is to prepare an accurate and complete flight log, just like the one's you haven't prepared since you were working on your instrument rating! It really doesn't take long at all if it's done as part of your regular preflight, particularly if you have one of the increasingly intelligent electronic calculators that are replacing the venerable "whizz wheels" of yore (although they aren't nearly as good as the slide of the old E6-B for scraping ice off aircraft surfaces!). Various pre-printed forms are available to keep things organized. The one I like best is available from several sources, and includes a little map of the United States on which you can sketch in fronts and weather systems and boxes for current and forecasted weather at your destination and alternate courses, as well as for clearances and the like. It also includes the usual grid of little spaces for courses, distances, speeds, times, fuel burns, and so forth. If you can get hold of one of these gems, Xerox off a bunch, or just buy a pad of them at your local pilot supply house.

The values of such a log are many. In addition to reducing the usual drifts of paper that accumulate in the cockpit on longer flights, the log warns you right away if flight conditions change, like for example, if the headwind picks up stronger than was forecasted (as it always does, at least when I'm flying!). It's also useful to review the flight log at the end of the flight, and be sure to save it if it's a flight you will make frequently!

Organizing Your Nav Materials

The next step is to get your navigational materials organized. Not that many airplanes have the room for those neat little racks that display the gleaming spines of your entire set of Jepp manuals, and even in those that do, that formerly spacious flight deck suddenly shrinks when you find that you have to have three or four binders open at once to get all the information you need! I've found it much more convenient to order a single extra binder from Jepp, with a few of the plastic chart pockets included, and make up a *trip kit* for each particular flight. I use one of the small leather binders, and it's more than large enough to hold the following, in order:

—SID(s) and primary approach for the departure airport.
—Area chart for the departure area, if applicable.
—Enroute chart(s) for the flight to the primary destination.
—Area chart for the destination, if applicable.
—STAR's and all approaches for the destination.
—Enroute(s) for flight from primary to alternative, if different from (3).
—Area chart for the alternate, if applicable.
—STAR's and all approaches for the alternate.
—In the pocket at the rear of the binder: Jepp plotter, a few spare pencils, and WAC charts covering the area of the flight—you never know when you might find a VFR chart darned useful.

Other Valuable Items

I also carry a lighted kneeboard for night flights, despite the faintly ridiculous "we're off to bomb Berlin" aura such gadgets always seem to have. Why? Because it lights up the approach plates better than most of the aircraft approach plates I've seen, and it can be strapped to the center of the yoke without a bunch of special adapters. Moreover, it replaces one of the two flashlights you should have at night, and in a way it's even more useful—if the instrument lights go out, you can strap the kneeboard to your leg backwards, and the light will shine on the panel without it having to be held.

There's one additional piece of equipment that I consider essential for single-pilot IFR or night operations; in fact, it's of so much value that it's required by law in most European countries (and in this one if you're flying the single-pilot version of the Cessna Citation jet). What I'm speaking about is some form of hands-free microphone, whether it's an aircraft-mounted boom mike, one of the little "stick it in your ear" earsets with the little tube-style microphone, or the full-scale "earmuffs" with lip mike is up to you, but in any case, it's worth its weight in gold. (As a matter of fact, up until early 1980, the tiny, lightweight earsets were worth their weight in gold!)

The value of such a mike goes far beyond its mere convenience—bending down and turning your head while searching for a dropped mike is virtually asking for a good case of vertigo, and the general contrast between merely flicking one thumb and speaking, or having to drop whatever you're doing to find the darn mike and answer Center before they get annoyed, goes far to promote the feeling of calm command that's so desirable.

Personally, I prefer the "earmuff" style headset, or at least the earset type, to the fixed boom mike, since the boom requires that you maneuver your head to it every time you want to speak. With a head-mounted unit, you can be looking at the panel, out the side window, or even at a chart and still be able to talk. I also am heavily in favor of the use of some sort of headphone, rather than the overhead cockpit speaker. Not only is the audio quality of headphones better and more easily understandable, but I feel that the passengers of an executive or even commuter aircraft have better things to listen to (such as their own conversation), rather than the usual incomprehensible blather that comes over the radio circuits.

Flying By Yourself

Very well, now that you're suitably prepared with precalculated logs, and festooned with kneeboards and headsets, let's get down to the business of flying. Again, remember that our objective is to stay ahead of the game.

We can start staying ahead when we copy our clearance. It's remark able how many otherwise competent instrument pilots fall hopelessly behind as the clearance is dictated, especially when you consider that any competent pilot, having already prepared and filed the flight plan, *has a pretty good idea in advance what the clearance is going to be.* You can use whatever shorthand you want, but I've seen pilots get bogged down copying, "ATC clears Navajo 59984 to the Santa Barbara airport," even though the pilot could have written *that* part down before even calling Clearance Delivery. If you look at your charts for a minute, you'll already have advance knowledge of the departure you'll be given, and most likely the departure control frequency as well. Given the current ATC situation, you may as well file the preferred route structure, if applicable (you're going to get it anyway)! That way, all you have to copy is "cleared as filed" and the squawk.

Next step (and you'd be surprised at how many experienced pilots *still* don't do this)—get your ducks in a row for departure: radios tuned, OBS's set, departure and/or area and/or, enroute chart out and folded to the proper panel, and so forth.

Now that we've taxied out and have completed our runup; it's time to go. As we lift off, we can start our elapsed time clock and note our off time on the appropriate spot in the log. It sounds trivial, but it's something that can be handy to know later on.

The next suggestion, be it ever so humble, is—*write it down!* This may sound trivial, but it takes only a second to jot down every heading, every new altitude, every amended clearance, every routing change, every holding instruction and every expected further clearance time; and it can go far toward eliminating the "say again syndrome." Again, it's a big contributor toward relaxation and peace of mind, and the more relaxed you are, the better job you'll be doing. You can use a separate pad, or a sheet of paper. My flight log has a big open space for such things, and I generally end up with a column of headings with all but the last one scratched out. I also include a similar column for altitude assignments, and a wider space below these for holding instructions and amended clearances. If you want to be fancy, you can get a little gadget that has adjustable thumbwheels to remind you of headings and altitudes (similar gadgets are built into the yokes of some airliners), but I've usually found that setting the thumbwheels takes longer than writing down the information. Of course, if your airplane has an altitude alerter system, you can also use that as a reminder. One Lear pilot of my acquaintance uses one of those little gray plastic "magic slates" sold in toy stores for the pre-school set. I don't know what this says about the mentality required to fly a Lear, but I found the little slate so handy that I bought one myself the next day!

As you cruise along, don't just sit there staring at the dials, look ahead along your course line on the chart, and keep tabs on where you are. Too many pilots have been lulled into somnolence by the womblike security of dual VOR's, DME, and the radar environment, so that when an FAA fuse blows somewhere and that little voice says, "Radar service terminated, contact Bigtown approach when you're 35 east of Rising Gorge VOR," they're completely unaware of their actual position. Try to anticipate all the horrible things ATC might feel like doing to you, and they'll suddenly become far less horrible.

If the weather becomes really nasty, it's even more important to stay on top of the situation (not to mention the ice-laden clouds). If you decide you need to deviate or go to the alternate, the earlier you know it and tell Center the easier it will be for both of you. If your radar shows something unpleasant ahead, don't wait until you've painted yourself into a corner; if you do, you're making it harder for yourself and the controller.

This becomes even more important during the descent and approach phases. You should get so far ahead of the situation, that you're mentally having a cup of coffee in the FBO lounge by the time you actually start down from cruise altitude. We touched on the importance of descent planning for recip twins in the chapter on reciprocating engines. You'll be busy enough on even the most normal of instrument approaches without the added burden of arriving "high and hot" over the initial approach fix, or worse yet, the final one!

Here's another thought: How often have you actually shot an approach right down to the minimums? And, more to the point, how often have you had to execute the missed approach procedure? I've ridden with far too may pilots who display a typical problem executing the miss—they should be looking at the instruments, ensuring adequate climb performance, but instead they have their head down on the approach plate, looking at the instructions for the missed approach procedure! For some reason, the prospect of a missed approach seems so remote to many pilots that they never look at the procedure until they're halfway into it! A much better, if somewhat pessimistic, philosophy is to expect that you'll miss every time, then you can merely enjoy the pleasant surprise when it turns out you can land after all. (A similar philosophy regarding engine failures on takeoff can be a real lifesaver, as we discussed at the beginning of this book.)

Being well-prepared for single-pilot IFR operations, then, is a matter of doing as much as you can when there's no pressure on you, allowing you to devote more attention to the situation when the pressure *is* on. In the next chapter we'll take a look at some of the severe weather you may encounter while flying an advanced twin.

8

Special Weather Environments

Flying larger and more capable aircraft will inevitably bring you into contact with a certain amount of "larger and more capable weather" sooner or later during your career. I don't mean to imply that the increased capabilities of your airplane mean that you can take anything you want, but the fact remains that you *can* sometimes consider flights on days that may well have kept you on the ground in lighter equipment. Moreover, it may well be that the mission you're engaged in, be it freight, commuter, or executive/corporate flying, will dictate flights on days when a more casual pilot would be just as content to sit it out.

Finally, the more advanced aircraft that you may be flying have the capabilities to fly in areas in which you may encounter weather that you'll find new and challenging. In particular, the higher altitude capabilities of turbocharged recips and turboprops, along with the fact that their normal operations require high altitudes for fuel efficiency on many flights, will lift you into a region of high-level winds, and, occasionally, into contact with that remarkable phenomenon—the *Jet Stream* (Fig. 8-1).

High Level Winds

The subject of high-level winds in general can be dealt with fairly briefly. Generally, at least within the performance limits of the airplanes we're dealing with here, the higher you go, the stronger the blow. Like the winds in the lower levels, the upper ones tend to veer clockwise with increasing altitude.

One good way to get an idea of the upper winds, and one that will give you a much better overall picture than the often-shaky FT forecast, is a look at the higher-level weather charts at your friendly neighborhood FSS. Up until now, you've probably been dealing primarily with the 850-millibar chart, corresponding roughly to an altitude of 5000 ft MSL, and the 700-mb chart, which represents the approximate 10,000-foot level. Now, though, you'll be looking at the 500-mb chart for 18,000 feet, and the 300-mb one for 30,000 or so.

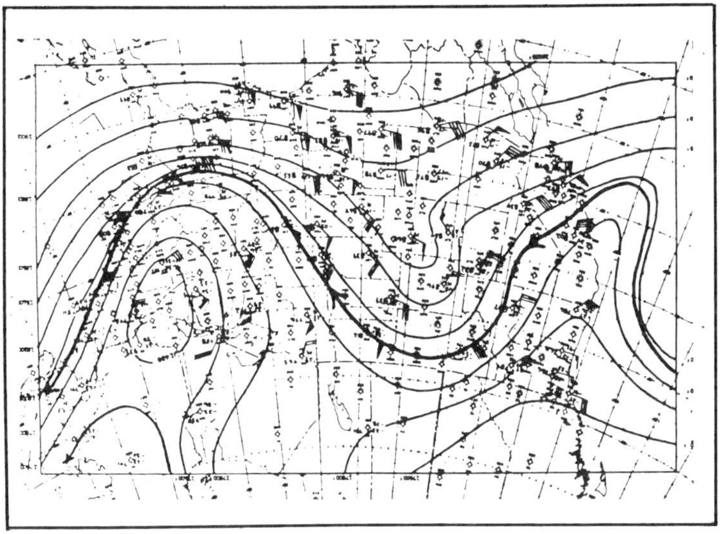

Fig. 8-1. On this 500-mb weather chart, the jetstream is shown by the heavy black line with arrowheads. Careful flight planning can make maximum use of its benefits, or at least minimize its adverse effects. On the surface wind arrows, each full-size "feather" is 10 knots; a "flag" is 50 knots (courtesy of USAF).

It's particularly interesting to compare the charts for a given day (Fig. 8-2). At lower altitudes, the contours representing equal height of the given pressure surface tend to curve quite sharply around highs and lows, thus the atmosphere is relatively "lumpy." At higher levels, the contours smooth out considerably, and often are further apart. The slope that does exist in the pressure surface is more gradual. As a result, the upper-level winds tend to be more constant over a larger area than at lower levels.

Be sure that you're looking at the correct sort of line on such a chart. The one of interest is the basic contour or height line. The winds blow along the lines with the lower *heights* (and hence lower pressures) to its left, and higher heights and pressures to its right (clockwise around highs, counterclockwise around lows, just like on the surface; in the southern hemisphere, it's reversed).

Be careful not to be confused by lines of equal wind speed, known as *isotachs*. They usually appear only on higher-level charts, and each line represents a line of equal wind *speed, not* direction! In fact, the isotach often lies nearly perpendicular to the actual direction of the wind.

The polar regions of the earth are covered with a mass of cold, dense air (Fig. 8-3). When the air mass encounters the warmer air that results from solar heating in the earth's equatorial regions, a *polar front* forms. Several other factors are associated with the polar front—the *tropopause* (the "dividing line" between the troposphere and the stratosphere) is lower over the polar air mass than over the tropical air mass, while the

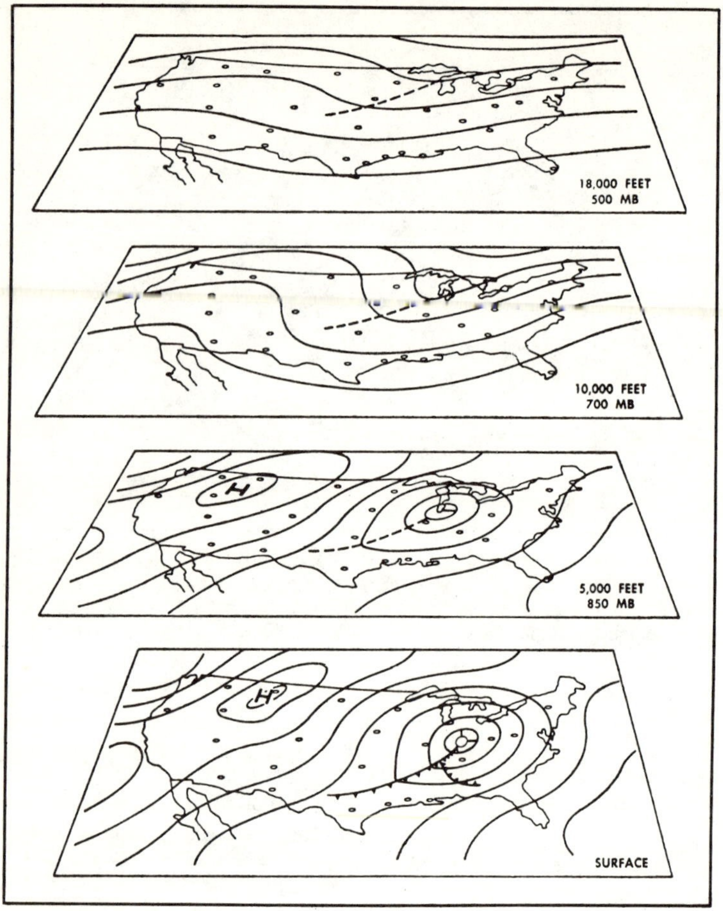

Fig. 8-2. Comparing weather charts for increasing altitudes, you'll notice that fronts and systems become less sharply defined, and changes in wind speed and direction become more moderate, at the higher levels (courtesy of USAF).

location of the front itself is further south at the surface than aloft (Fig. 8-4). The upper limit of the polar front is roughly below the discontinuity in the tropopause, and this is the type of region in which the jet stream can form. It's a relatively narrow "tube" or "river" of air aloft, flowing along the polar front from west to east, and it can reach velocities of up to 200 knots!

While the jet stream core itself generally is found at altitudes in the low to mid thirties, and is hence somewhat above our normal cruising altitudes, it can entrain air in the region surrounding it and cause winds as high as 100 knots or more. These are good ones to be aware of, either to "hitch a free ride" if you're going in their general direction, or to avoid them if you're not. Such winds can be found at altitudes as low as the mid-to-high teens and low twenties, and, while the jet streams sounds like a polar phenomenon, bear in mind that in winter months it frequently dips as far south as 20° latitude.

Where can you find the jet stream, and how do you go about using it? Well, its core is usually some 200 nm behind the polar front (surface location), but useable winds will be found up to around 200 nm south of the jet stream (near the front itself), and slightly less to the north of the stream core, Higher wind velocities will be found at lower altitudes on the south side. You can spot the jet stream as a heavy line on the 500-mb chart. If you still can't find it, it should be almost parallel to the front. Thus, in the event of the very common frontal wave systems, you may well find the jet stream crossing the entire U.S.A. in a single immense S-curve.

Some clear-air turbulence is often associated with the jet stream as it "scrapes along" the slower-moving air around it (Fig. 8-5). This turbulence is generally restricted to a vertical band only 1500-2000 feet thick, so if you're ever climbing eastbound to high altitude and start running into

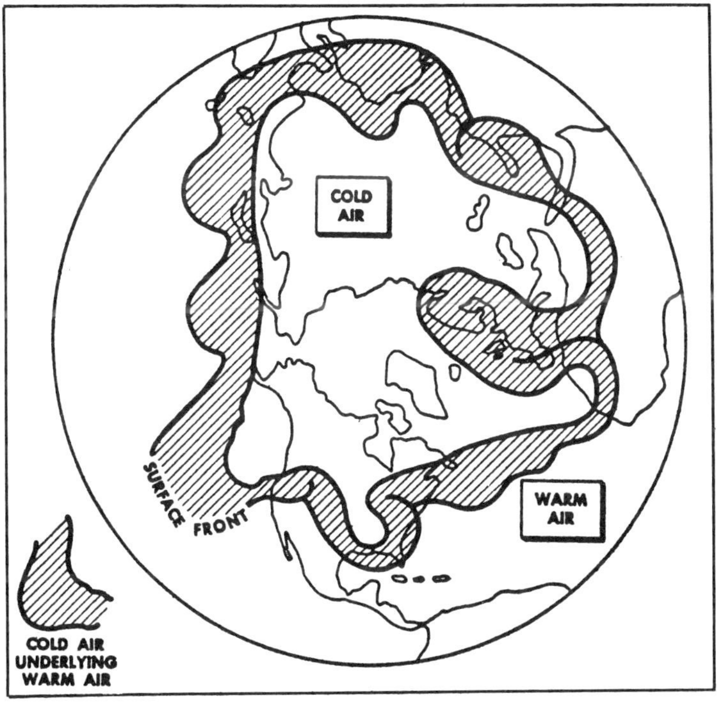

Fig. 8-3. Here's a view looking down at the north pole, showing the location and slope of the polar front. The scallops and waves in the surface front are what give birth to weather systems, moving from west to east (counter-clockwise) (courtesy of USAF).

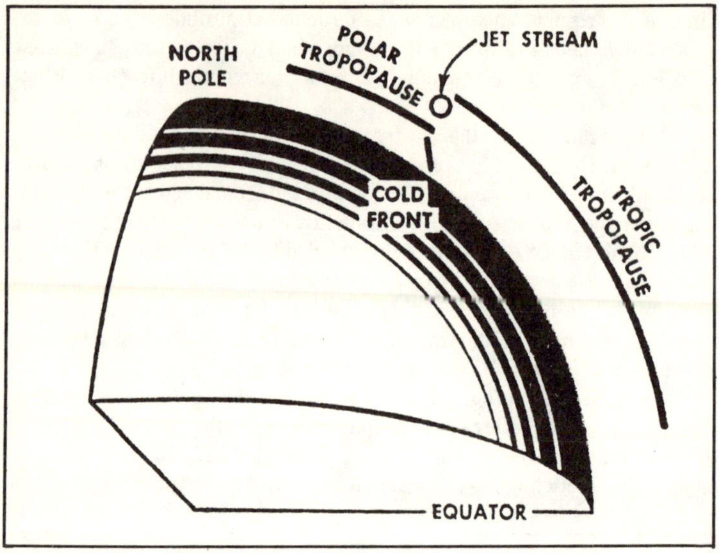

Fig. 8-4. The jet stream is found in the "gap" between the polar and tropic tropopause. Look for it somewhat north of the surface location of the polar front (courtesy of USAF).

unexplained chop, it may well be worth it to negotiate a couple of thousand extra feet with ATC, even if you didn't plan to use the jet stream (Fig. 8-6). It's an uncommon and unforgettable experience to feel the chop ease off and see the DME ground speed readout suddenly wind up an extra 120 knots. The first time it happened to me, I thought the DME was malfunctioning and turned it off! Sometimes the jet stream dips to relatively low altitudes. I know of one occasion in which a single-engine Cessna Centurion was flown nonstop from the Pacific Northwest to Pennsylvania in just a bit over seven hours!

What about those other times, when you reach cruise altitude and the DME indicates you've slowed to a crawl? If you suspect that you're in the jet stream, your best, make that "least worst", move is to turn northward and descend into the cold air mass behind the front, where the headwinds won't be so bad for so long. You'd still better settle back for a long day in the saddle, though!

Icing

Flying an advance twin may well represent your first real exposure to icing conditions. Why? Because for most lighter aircraft, forecast or known icing conditions mean just one thing—don't fly. The FAR's are pretty explicit, too: " . . . no person shall operate an aircraft under VFR in known conditions of light or moderate icing . . . or IFR in forecast on known moderate icing . . . without functioning de-ice equipment . . ."

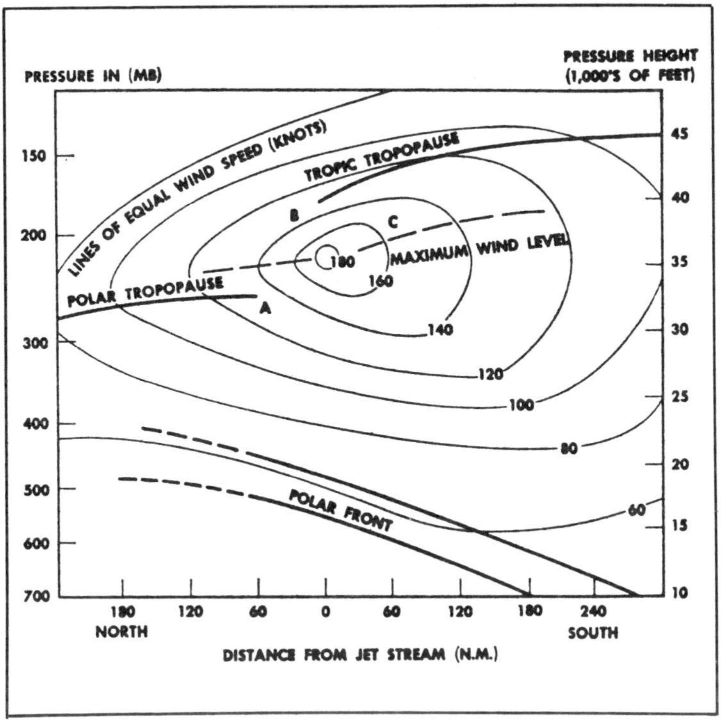

Fig. 8-5. If you were in the jetstream, looking downwind, this is what you'd see. Turbulence can be expected at points A, B, and C. Notice that even without getting into the jetstream itself, you can find healthy tailwinds at moderate altitudes, especially on the southern side (courtesy of USAF).

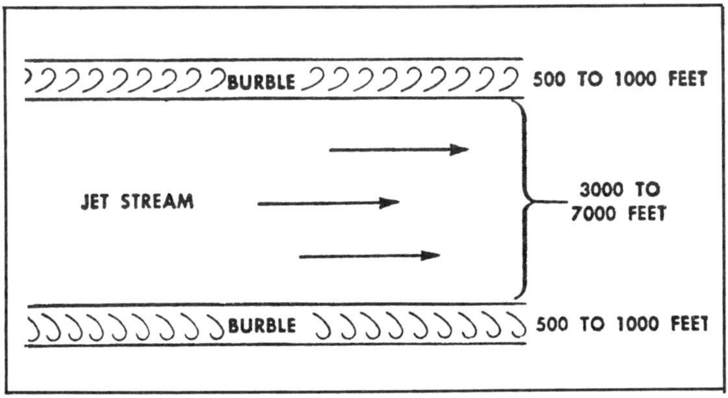

Fig. 8-6. Light to moderate turbulence can be expected both above and below the smooth air of the actual jetstream. Often, a change of only a few hundred feet of altitude can smooth out high-altitude clear-air bumps (courtesy of USAF).

One point that's a bit hard to clear up is the difference between "forecast" and "known" icing. When you call up the FSS and they give you the old refrain of "light to moderate icing in clouds and precipitation," it's likely to be part of the area forecast, so, at least legally, you could fire up and go VFR and hope to avoid the icing conditions. It wouldn't be a good idea, though! If, on the other hand, the briefer adds " . . . Cessna 421 reports moderate mixing icing climbing through 13,000 feet . . ." then you know its there. The point becomes even more moot when you consider that some aircraft are not certified for known icing conditions, even though they may well be decorated with all sorts of de-ice equipment.

We've already covered the types of equipment available; let's look now at the factors that actually cause icing, and what you can do to minimize the problems.

Like any other weather phenomenon, icing conditions are not absolutes, and, while there are some times when just about anything should stay on the ground, there are many occasions when a proper assessment of the conditions can combine with the right equipment so that icing need not compromise the utility or convenience of the light aircraft.

Attitudes about icing conditions vary widely. One camp suggests that to venture into icing conditions in anything less than a jet with hot wings is sheer suicide; on the other hand, a well-known article a couple of years ago by a highly-experienced Alaskan bush pilot suggested that ice isn't really as much of a problem as most people think, and can be dealt with quite simply. (The credibility of this latter article suffered, however, when its author and a VIP passenger disappeared on a flight in icing conditions only a couple of months after its publication.) It seems obvious that there is some middle ground on which both factions can meet and agree that, "Yes Virginia, ice is something to be treated with the greatest respect, but something that a well-prepared pilot and aircraft can handle safely."

Let's look at the conditions that cause an airplane to ice up. A quick peek into the freezer compartment of your refrigerator will reveal that there are only two conditions that have to be met: a sufficiently low temperature, and a supply of moisture.

Obviously, low temperature is available to anyone who's willing to climb a few thousand feet; exactly *how* low can be very important. Ice can form at temperatures ranging from a degree or two *above* freezing (on certain parts of the aircraft) to well below freezing—minus 20 degrees Centigrade is a good "ballpark" figure. As we will soon see, the worst temperatures aren't necessarily the coldest, but rather those just below freezing.

The moisture we need for ice formation is of a rather special sort called *supercooled* water. The processes involved in super-cooling are still not entirely understood; it is sufficient to say that in small droplets, water can remain liquid at temperatures well below freezing. These droplets, however, are in rather precarious equilibrium at best, and the slightest impact, for example, against the leading edge of a passing airplane, will upset this delicate balance and cause the droplets to freeze instantly.

Different conditions of temperature and humidity can affect the size and frequency of these droplets and, indirectly, the severity and rate of ice accretion on an aircraft. Cold air cannot hold as much moisture as warm air, so the clouds which form at low temperatures have a lower moisture content than those forming in warmer air. The supercooled droplets in such clouds tend to be quite small, and freeze instantly upon contact with an aircraft, giving rise to *rime* ice. Rime usually has an opaque, almost "milky" appearance, and it's relatively brittle and hence easy to dispose of with de-icer boots. Large amounts of air are trapped in the rime-ice freezing process, which not only causes the milky apperance but make the rime ice fairly light in weight; unfortunately, weight is a secondary consideration in icing problems. Ice's main effect on an airplane's performance comes, not from its weight, but from its additional drag and changes in the airfoil, and in this regard, rime can be every bit as bad as clear ice.

Warmer air, on the other hand, can hold more moisture; the supercooled droplets are often larger. Their size, as well as the warmer ambient temperature, makes them feeeze more slowly, and gives them more time to "splatter" and run along the aircraft's surface. This is the condition that causes clear ice, which is not only much heavier but also much denser and tougher than rime ice, and is much harder to get rid of. Moreover, while rime ice accretions tend to build foreward into the airstream, clear ice tends to work backward, downstream, from its initial appearance.

Perhaps the worst ice of all is that encountered in situations where precipitation, usually rain, falls from a warmer area aloft into sub-freezing air below. Such "freezing ran" can be, in the words of one high-time airline pilot, dynamite. It can load an airplane with a paralyzing load of ice faster than its de-icers can dispose of it. In one classic instance, a DC-2, long esteemed as an airplane capable of carrying a load of ice that would make a Good Humor truck slink away in shame, ran into freezing rain some 10 minutes out on an approach. The pilot decided to make a run for it; despite the best efforts of the de-icers, the airplane could barely make it to the threshold with full power, and ground crews had to batter ice away with chisels and crowbars before the door could be opened.

One of the most effective ways to deal with ice is to avoid it; with an understanding of the factors involved in ice formation, this task becomes much easier. Even if icing areas can't be avoided altogether, an understanding of icing mechanisms can minimize the problems. The small droplets mentioned above, for example are often found in the status decks present after the passage of a front; pilots flying in such decks can expect rime ice, but can usually console themselves with the thought that these wintertime stratus layers are seldom more than a few thousand feet thick, and that a climb on top will usually end icing problems.

By the same token, the more serious clear ice is usually found in clouds of significant vertical development, where warmer and moister low-level air is carried aloft. This is often associated with turbulence. A pilot aloft in clouds above the freezing level who feels bumps, or who notices that the outside temperature is rising—should be alert for clear ice, or mixed clear and rime ice.

The worst icing occurs in or near frontal zones; both warm and cold fronts can give rise to icing conditions. This is where preflight planning can really make or break a flight—knowing which type of front one's dealing with, and from which side one's penetrating it, can make the difference between smooth sailing and tense flight wondering if the de-icers can stay ahead of the situation. Cold fronts are usually more intense, but narrower; warm ones often don't have weather quite as strong, but you have to sit and sweat it out longer. Knowing one's orientation can be all important if one runs into ice, for example, if you know you're penetrating the front from the cold to the warm side, you can be pretty sure of climbing into warmer air (if that's what you want to do—remember, if you're picking up light ice in really cold ir, it'll be heavier as you get nearer the freezing point). On the other hand, if you're coming from the warm to the cold side, a climb might end up right along the frontal slope, leaving you in the worst possible conditions for the longest possible time.

There are times, of course, when penetration of a front must be accomplished on a "let's try and see" basis, typically when one's coming up on the front IFR from the trailing side, and can't really see what's going on. You may know the approximate location of the front, but your first real indication may be turbulence and clear ice. This is the time when you'd better have a very definite idea of how much ice you're willing to carry before you're through—and your decision to make a 180 will have to be based on *half* that amount, since if you do turn around, you'll have to come out through the same icing you penetrated when you went in.

Dealing With Ice

Let's turn now to what we can do about ice, both in terms of technique and of equipment. Even in a plane equipped with de-icers, the best thing to do is get out. The first appearance of ice is the time to open negotiations with ATC for a better altitude or routing. Obviously, contact with an FSS in invaluable for reports on tops, icing conditions further along the route, frontal location, and so forth.

One's instinctive desire seems to be to be to descend. Often this is the best course, but there are a number of factors that should be considered. If the temperature is only slightly below freezing, and if rising terrain or lowering freezing levels (which often go hand in hand) are no problem further along, descent by all means. On the other hand, if the temperature is already pretty cold, perhaps a climb might be a better idea. The worst icing occurs right below freezing, and the situation gets progressively better the colder it gets; by –10 degrees Centigrade, the worst is over, and by –20 degrees ice is pretty rare. Thus, even if the tops are far above you, a climb may make a lot of sense, just be sure that you aren't climbing into warmer temperatures aloft that would still be below the freezing level (i.e., a winter warm front) or heading toward an area where storm activity will be carrying warm, wet air up to you.

If you do start a climb in icing conditions (by the way, don't haul the plane back ot a normal climb speed), use lots of power and make a flat, high-speed climb. Why? Because at high angles of attack (low speeds), more of the bottom of the airplane and the wing are exposed to the airstream that will collect ice you'll have no way of getting rid of.

Some aircraft, regardless of equipment, are better ice carriers than others. Size is one factor, of course; just because we hear some old-time airline captain talking about carrying four inches of ice on a DC-3 doesn't mean we can get away with it in a lightplane. It's not just the weight, although that's a factor; we also have to bear in mind that a four-inch contour change hurts a DC-3 wing, with a chord of something like 15 feet and a thickenss of well over a foot, far less than it would a lightplane wing with a chord of four feet and a thickness of seven inches or less.

"Cleanness" is another point. The more gadgets sticking out into the airstream—antennas, struts, landing gears, and such—the more places there are for ice to collect and build up drag and weight.

There can be another problem with this, too. Antennas, especially the wire whip type, can vibrate when they pick up asymmetric loads of ice. Not only will they set up an ungodly howl, but it can get to the point that they carry away altogether, leaving you without communication at a rather bad time. The plastic-covered blade or spike type are better, although their electronic performance is still adversely affected. Other airframe parts can also vibrate and rattle. Sometimes in rime conditions, the airflow will alternate between the sides of an obstruction and will build up large ice *butterflies* or *bow ties* extending several inches to each side, with attendant drag increasing.

The cautious pilot will only enter icing conditions with a considerable performance margin. If the icing is to be expected in a status layer, for example, with clear air on top, the ideal airplane is one that can climb rapidly through icing layer and be on top before too much ice has collected, one of the times when a turbocharger can be extremely valuable. By the same token, the airplane should be able to fly adequately with a certain amount of ice aboard. Every year, we read about icing accidents in the Western mountains where light airplanes have gone down with relatively small amounts of ice, since they were operating close to gross weight, and to their service ceilings.

Making Your Gear More Effective

There are a few things you can do to increase the effectiveness of the gear you have. A big problem is when hunks of ice stick to the boots and go up and down with them. I can recall one instance, with a mixed load of clear and rime in a 310, when the first try with the boots broke off all the nice smooth rime ice, leaving a bed of clear ice that looked like broken bottles. Airspeed dropped by 8 kt and stayed that way until we could let down into warmer air. Various silicone-spray preparations, such as *ICG*, are available, spraying some on the boots before an icing flight seems to help quite a bit. It's good stuff for propellers, too, many pilots without hot props swear by it. (It washes off in rain, though.)

Jets get around the problem somewhat differently, since they have almost unlimited quantities of hot compressor bleed air available. The air in simply ducted through engine colw lips, inlet guide vanes, and wing and tail leading edges to heat them above freezing. In fact, even a few late reciprocating transports, the Boeing 377 Stratocruiser, for example, had wings heated by gasoline combustion heaters built right into the leading edges. This is another of this systems that can be used either for de-ice or anti-ice. Again, the latter is preferred, since if the heat is turned on with ice already on the wing, it'll just melt and run back until it's past the heated section, where it refreezes on a rather critical part part of the airfoil.

It goes without saying that the airplane will fly quite differently if you do pick up a good load of ice. It'll take more power to maintain airspeed, and the stall speed may be much higher than normal, due both to the weight of the ice, which can amount to several hundred pounds even on a relatively small aircraft. The increased weight can lead to increased drag and changed airfoil shape. It's quite likely that the center of gravity will have moved, too. All of these factors are cumulative, so an iced-up approach and landing, not all that unlikely in the event of encountering icing you're not able or willing to handle, should be flown at speeds well higher than normal, and the high speed should be maintained all the way to touchdown. (Since the conditions that cause heavy icing may well also cause icy runways, this can give rise to some rather tense moments.) Moderate and heavy icing is, as we said earlier, not the place to see how well your de-ice equipment really works. Once the ship's well and truly iced up, the rate of further icing will go up, since the load already aboard offers more surface for new ice to grow on. You should have contracted ATC long before this for a better altitude or routing. Remember that the final responsibility for safe flight is yours, and if ATC holds you in icing conditions, make the gravity of the situation clear to them. In extreme situations, it would be better to start a descent and argue later than to stay that extra minute and pick up so much ice you have no choice but to descend anyway.

In the final analysis, safe flight in icing conditions depend on careful and complete preparation of both the machine and the pilot. The airplane might be anything from a 150 with a hot pitot tube to an ailiner, depending on conditions. What's important is that it should be in top shape throughout, so don't check just the de-ice equipment. Buzzing along in light rime is not the best time for, say the OSB knob to fall off in your hand; you've got enough to deal with as is. Carry all excess fuel you can, because you might find yourself operating at a much higher power setting than you planned.

Be Prepared

Pilot preparation's a bit more complex, and should start months before the flight: the better you're armed with basic knowledge, the better you'll be able to deal with new situations. The FAA's *Aviation Weather* handbook (available from the GPO and at many FBO's and flight schools) contains a lot of the basic information.

Armed with the basic weather and icing knowledge, you can make much better use of the weather information available to you before a flight. If you're contemplating a trip that may involve ice, get all the information you can on such things as the freezing levels and lapse rate. This will allow you to know how far you'll have to let down to be safe or climb to be too cold for icing, as well as the moisture content of cloud layers enroute, which will tell you whether to expect clear or rime ice. Know where the fronts are, and which way they're moving, since that's where you can expect the strongest icing, and make sure that you always have an escape—warmer air above or below, clear air between layers, or a really solid enroute alternate with runways long enough to cope with an iced-up "warp nine" approach.

All of this notwithstanding, there will still be times when discretion will not only be the better part, but in fact the only part of valor. But, there will also be times when a properly prepared pilot in a properly prepared aircraft, armed with adequate knowledge of the situation as well as his own and his airplane's limitations, can reach his destination with a minimum of difficulty and delay even on those days when the teletype is printing out, "LGT/MDT ICGICIP."

Thunderstorms and Turbulence

Regardless of the size and equipment of the airplane you're flying, there's a basic rule for the preparation of thunderstorms: *don't*.

Flying an advanced twin, though, you're more likely to be aloft on days when thunderstorms are about, and especially in warm-front weather, you may blunder into one. A bit later in this chapter, we'll offer a few suggestions for this unhappy state of affairs. For the moment, though let's turn our attention to *avoiding*, rather than penetrating, thunderstorms. Knowledge of how, where, and when they form is already a major step in the right direction, and we'll touch on those areas in a moment, but many of today's advanced twins offer their pilots a really excellent electronic aid. Yes, when those cu-nims build up and the stability index drops, we spell "relief" R-A-D-A-R.

Understanding Your Radar

Let's get one thing straight at the outset. Having a radar set, whether it's one of the old "smear-o-matic" tubes or a more modern digital type, *does not allow you to penetrate thunderstorms safely!* In fact, on some days, it may well be the radar-equipped airplanes that'll turn back first, to land in comfort and wait it out, while the non-radar aircraft forge ahead and get bombarded.

What you can penetrate with radar is a *region* of thunderstorms, by picking your way *between* them. This is assuming, of course, that you have some knowledge about the causes and behavior of thunderstorms, and about the limitations of even the most sophisticated radar systems.

Let's examine those limitations. What we're trying to avoid in a thunderstorm environment is turbulence and hail. Can the radar see turbulence? No. Can it see hail? Only very faintly.

What radar can see is water—*liquid* water, in the form of rain. It sees big raindrops better than little ones, and hard rain better than light rain. It can see hail, but not very well; a hailstone's reflectivity to radar beams is only a small fraction of that of rain. Thus, it's a good idea to know *where* in a storm system hail can be expected; otherwise you may fly into what you think is light rain, only to lose paint or even a randome or windshield.

The whole idea of airborne weather *avoidance* radar is based on the (sometimes tenuous) correlation between rainfall intensity and turbulence. Simply stated, the heavier the rain, the more likely it is to be rough. There's another factor, which has only become well-known since the introduction of "contour" circuits on aircraft radars—the steeper the *rainfall gradient*, that is, the shorter the distance over which the rain intensity changes from light to heavy, the rougher the associated turbulence. The newer digital radars, particularly those with color display, make avoidance of such areas much easier than before.

The function of the *contour* circuit on a radar set is to blank out all rainfall *stronger* than a preset rate. Thus, on a non-color indicator, the strongest rainfall zones within a storm will be indicated as black holes, rather than by light. To check whether such an area is, indeed, a real hole or a contoured area, the contour circuit can be momentarily switched off (some radars do this automatically on alternate sweeps). If the "hole" fills in, it's not a hole, it's a hard spot; if it stays dark, it really is a hole.

General Rules of Radar

A complete discussion on how to use radar is beyond the scope of this book. You can check the manual supplied with your radar set. An even better idea is to attend one of the excellent radar schools presented by United Airlines. They usually offer them at certain gatherings of the clan, such as the NBAA show or the Reading, (PA) airshow, or you can contact them about attending a regular session at their Denver, CO training base. There are a few general rules to bear in mind, though, and we'll go over them now.

First: remember that more distant storms may well be hidden behind nearer ones. Radar can "see through" light rain, but not nearly as well through heavy stuff. If you're using radar to pick your way through an area of strong returns, be a bit cautious that it's not leading up a blind alley into worse stuff on the other side.

Second: since radar performance deteriorates at longer ranges, be particularly cautious about anything that it spots off in the distance. A storm big enough to show up at all at 100 miles will probably turn out to be a real bad one when you get closer, so give anything you spot at the longer range scales plenty of room.

Third: watch out for sharp gradients. A small area of contour surrounded by a narrow band of lower intensity rainfall is likely to be a lot worse than an area surrounded by a large region of lesser rainfall. Remember, the steeper the gradient, the rougher the ride.

Fourth: avoid like the plague any of the following: *scalloped edges; projecting "fingers;" hook shapes* (sometimes *resembling the figure 6 or 9; and spearhead or arrowhead shaped storm areas*: The scalloped edges and hooks are a warning of likely tornado or funnel cloud activity, as are the fingers to a slightly lesser extent. Also, the finger regions, which may be separated from the main body of the storm by a considerable distance, contain *gust tubes* leading back into the storm at very high velocities. Spearhead or arrowhead echoes have been found to be associated with particularly violent descending air currents called "down-bursts," which can cause massive wind shear at or near the surface.

Fifth: use the radar's tilt control for the best effectiveness. Remember that the radar can see rain much better than ice and snow, so if you're flying at or above the freezing level, you'll get a better idea of the location of the active cells by tilting the antenna downward a bit, particularly at closer ranges.

Penetrating a Thunderstorm

The time may come when you'll have to penetrate an area of thunderstorms. Hopefully, you'll be forewarned, either by radar or by careful preflight scrutiny of the weather picture. If you're coming up on a front, you'll know what kind it is and from which side you're approaching it.

To some extent, your tactics for dealing with a line of thunderstorms depends on such knowledge. For the purposes of this discussion, we'll assume that we're dealing with a front or a prefrontal squall line (which, incidentally, is where you'll find the nastiest of all thunderbumpers). Air-mass thunderstorms, such as the ones that form in the West or Midwest on a warm, humid afternoon, are generally scattered, and you can pick your way around them either visually or with radar. Not so with the frontal type, which may stretch in a line from horizon to horizon; worse yet, especially with warm fronts, you may be in IMC and unable to see them until you're in among them. What's the "least worst" technique for getting through with a minimum of discomfort?

One important factor, given the ATC situation, is picking the right altitude for the conditions. Cold-front thunderstorms and squall lines (Fig. 8-7) generally extend close to the ground, and have nasty weather in their lower regions, so trying to stay low would be a bad choice. Warm-front

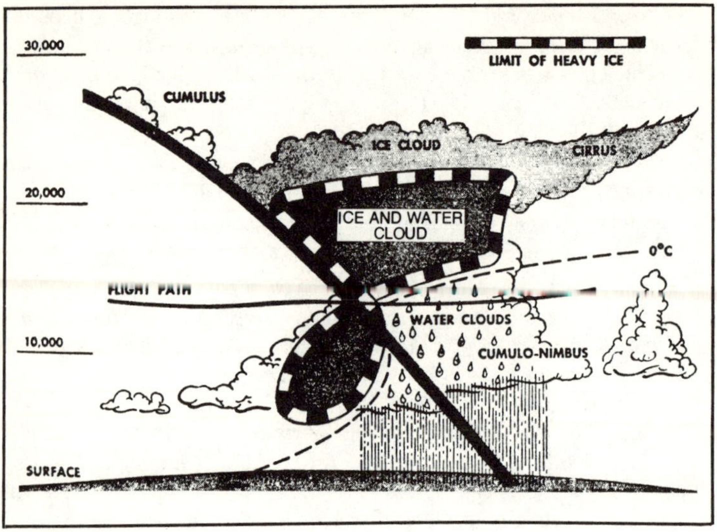

Fig. 8-7. In a cold-front thunderstorm or squall line, the area of heavy icing is fairly limited in size. Due to the steep slope of the frontal surface, minimum icing exposure would result from a flight path crossing the front right at the freezing level (courtesy of USAF).

types (Fig. 8-8) on the other hand, tend to leave a bit more room underneath. Sometimes, in fact, descending to the MEA will keep you below the worst. It may, in fact, be a good idea to cancel IFR if there's sufficient ceiling below the (usually stratus or stratocumulus) clouds, since you can fly VFR and avoid the areas of heaviest precipitation, which will be near the worst cells.

A word of caution, however: should you observe the so-called *Cumulus mammatus* formations, stay clear of them. These are clouds which puff *downward* from a stratus layer, shaped rather like a big egg. Remember that they're a frequent precursor of tornadoes and violent downbursts.

Cold-front types offer you fewer options. If you have the altitude capability, you might be tempted to go through a low spot in the line of storms. But, be aware that those low spots can fill in with stunning quickness, and if they suddenly become an area that's rising rapidly, they're more than likely to harbor active cells. If you're lucky, the freezing level will be high enough so that you can stay well below it, which isn't a bad idea for a couple of reasons. Often, some of the heftiest bumps occur right around freezing—and, of course, just above freezing is where you may encounter snow, which is OK, and hail, which isn't. Finally, most electrical discharges from cloud to plane (we can't really call them lightning strikes in most cases) seem to occur in the region from about ten degrees above freezing to ten degrees below it.

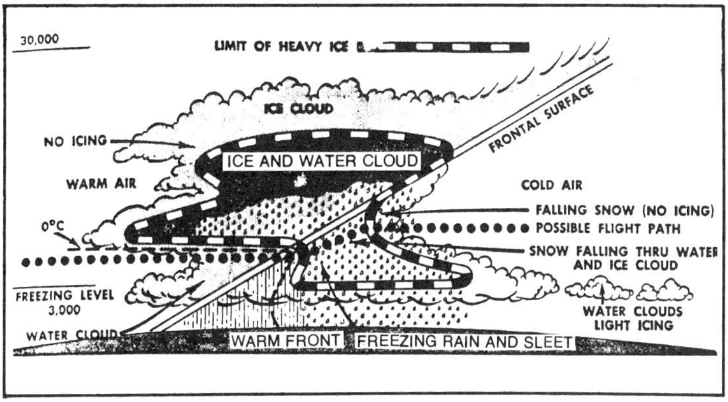

Fig. 8-8. Because of the relatively shallow slope of the frontal surface, warm-front storms tend to contain larger areas of icing conditions. Be aware of your location relative to the front to avoid prolonged climbs parallel to the frontal slope (courtesy of USAF).

A final warning before we plunge into the murk: if you're going to try to stay in the clear but pass fairly close to a storm, be very wary of going under the "anvil" at its top. A lot of hail is formed up there, and most of it falls *outside* the main body of the cloud—right on airplanes passing below the anvil. Why not circumnavigate the storm on its upward side instead?

All right, we've determined what sort of front we're going to have to cross (Fig. 8-9), and we're about ready to head into it. First, let's get our craft rigged for heavy weather. This means *everyone's* seat belt should be fastened, and not just on, but *tight*. Attach the shoulder harness, too, for

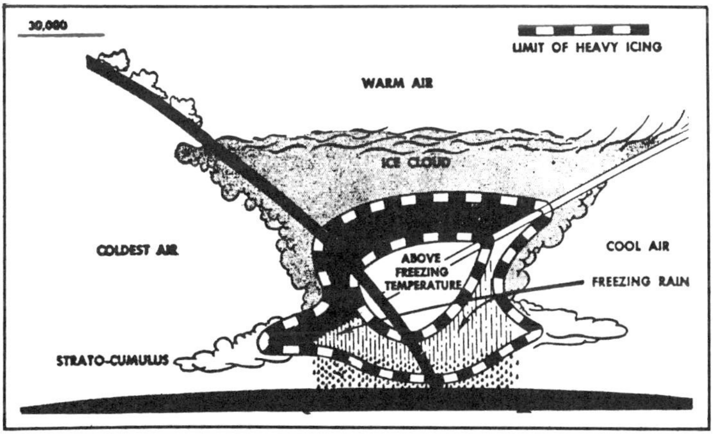

Fig. 8-9. Occluded fronts can combine the worst features of both cold and warm front systems. Minimum icing can be expected from a flight path at or near the altitude of occlusion, a difficult item to forecast! Higher altitudes will keep you clear of the freezing-rain zone (courtesy of USAF).

every seat that has it. Loose objects should be stowed; all you need in the middle of a thunderstorm is for a Jeep Case or flashlight to massage you upside the head.

Next, once in the clouds, we're going to be on the gauges anyway, so let's lower our set to the bottom of its travel to limit how much we'll see of any lightning flashes. Then, we'll turn on all the cockpit lights as bright as they'll go. Most civil aircraft don't have the really bright thunderstorm lights that military ships have, but even the cockpit overhead ceiling light (white) is a big help. Turn on the passenger's lights, too, it'll make them look apprehensive. (If the ship's big enough to have a bar, you can warn them that drinks and the like may be a bit sloppy in the turbulence, but if they're apprehensive you might want to leave them enough time for a quick nip before you enter the storm!)

Turn on the pilot heat, since the freezing level within the storm may change widely over only a mile or so. Follow the same procedure for props and the windshield. If you're flying a turboprop, engine inlet ice protection is particularly important in the heavy rain and hail environment of the thunderstorm.

Finally, if VFR, pick a low spot in the upper clouds to go in under, but watch the precipitation: Look on the radar screen—sometimes darker is better, since dark means rain, while while can mean hail. Ready? Here we go!

If you're crossing a cold front, starting at the warm side, you may find that the worst bumps will hit you pretty soon, in fact, they may even come before the worst of the precipitation. The key word here is *attitude*. Before entering the storm, you'll already have slowed the airplane to its turbulence-penetration speed; as you do, note the attitude that it takes to hold that speed in still air, and set the little airplane in the gyro horizon accordingly. From now on, that attitude (and wings level) will take precedence over airspeed and altitude.

Autopilots and Bad Weather

Perhaps we should take a moment to say a word about autopilots: if you have one, especially a modern one with lots of authority use it, but only as a basic autopilot, not in its coupled modes. That is, *don't* try to use altitude hold or airspeed hold, just set it for a level pitch attitude. By the same token, don't use any navigation modes, just wings-level; if it won't hold wings level except when heading mode is engaged, set it once for the heading through the storm and then leave it alone. Be ready to disengage it if the situation gets beyond its capabilities. Some units will disengage by themselves under certain G-loads, often without warning the pilot, so don't ever depend on it. You should just use it to help you.

Getting Into the Thick of Things

As you get further into the storm, the bumps will continue, sometimes as quick jolts, other times as big up-and-downdrifts that may change your altitude by hundreds of feet. Let them, it's much more important to hold pitch altitude. Chasing altitude and airspeed is generally what gets people in trouble in turbulence, and especially in thunderstorms. As long as the airspeed stays reasonably close to our chosen maneuvering speed, don't worry about fluctuations. If necessary, make power changes, but not overly large ones; if you do, trim to maintain the chosen pitch attitude.

Precipitation will begin and get stronger, ranging from light rain to firehose intensity. Unpressurized airplanes will often reveal leaks that you never suspected before; just grin and bear it! As long as it's just rain, all it'll do is wash the airplane for you.

What about hail? Nasty though it sounds, grinning and bearing it may be the best bet here, too. Anytime you try to maneuver in hail, you're extending control surfaces from their normal streamlined position, thus exposing them to damage. Turning the plane around will expose a whole side to damage, and cracking a window in a pressurized airplane while you already have your hands full is no joke. Areas of hail are generally very localized, and you'll probably spend less time in them by just gritting your teeth and pressing on than if you try turning to avoid them.

Then, there's the question of electrical disharges, so-called *lightning strikes*. True, lightning strikes can only occur between the clouds and the ground, or the clouds and other clouds, but when something zaps your airplane, who wants to exchange words?

Such electrical discharges occur when the airplane builds up a charge by flying through precipitation. Often, they're presaged by lower-level displays, such as *St. Elmo's Fire* which is an eerie blow glow visible at night on prop tips, radio antennas, and the like. Sometimes, you may hear a rising howl or squeal even on VFR radios just before a discharge, or your hair may prickle and stand on end (if it isn't already)! Occasionally, you can discharge the accumulation by keying a mike, at which the air around the airplane may seem to flash for an instant.

Not one pilot in a thousand has really experienced a true discharge. When they do occur, the usual manifestation is a bright flash and a loud pop, nowhere near as loud as a clap of thunder, though. Sometimes a discharge won't leave a trace. Other times it may burn a small hole, say about the size of a quarter, in some relatively pointy part of the airplane, line a wingtip or the end of the tail cone. Metal structures seem to be relatively invulnerable, since they can drain away the electrical charge. Insulation structures, like fiberglass tail stingers and the like, don't fare quite so well.

If it ever *does* happen to you, feel honored. Apart from that, there's not much you can do, although it wouldn't be a bad idea to regard the "wet" standby compass with a certain amount of suspicion. Some of the more advanced and sensitive modern avionics units may also indicate *tilt* on their little digital displays. In many cases, this can be cleared by turning them off and on again.

As you come out the back of the storm, the precipitation will decrease, as will the turbulence. Don't relax completely until you've put ten of fifteen miles behind you, but the worst is over.

If you'd been coming from the other side of the front, the sequence of events might have been more or less reversed, and this is something that can sometimes be embarrassing, to say the least. You'll have hit the precipitation first, and then the worst of the turbulence. Sometimes, in fact, you might fly right out of the storm (on the "warm" side) only to have it deliver its Sunday punch of turbulence in relatively clear air.

Let me reiterate one more time: the best way to penetrate thunderstorms is not at all. There's one other way to "penetrate" them that's a good one—on the ground.

How can you do this? Well, consider that the front is moving at, say 20 to 30 knots. Obviously, if you're coming up behind it, there's nothing to do. If, on the other hand, it's coming toward you, it's an easy matter to land sufficiently far ahead of it to avoid its downburst gusts near the airport, then enjoy a cup of coffee on the ground while the storm moves over you. You've just "penetrated" it at 30 knots or so, while sitting down in comfort. This doesn't sound like the best use of your expensive and sophisticated airplane, but think about it for a moment—if you were already at a fairly low altitude, you'll probably lose only half an hour or so. That half an hour would have seemed a lot longer if you'd spent even part of it penetrating a region of thunderstorms.

Approaches in Low Visibility

In a sense, this is an area where you're moving into more critical aircraft. While the larger twins are better equipped and more stable, they're also faster on final approach. In fact, they may well be in a higher approach category than the light twins you learned in, and so (at least theoretically) you'll be shooting approaches to higher minimums.

Unfortunately, in real life it doesn't work out that way. First of all, your overall increased weather capability may well have you out flying in worse weather than before. Increased aircraft equipment may even lead to your seeking and gaining authorization for Category II ILS approaches. Moreover, those higher minimums for bigger airplanes are there for a reason: 200 feet and a ½ mile go by a lot faster at 120 knots than at 80! A bigger airplane also needs more time and space to transition from the approach phase to the flareout and touchdown. All these factors combine to require good technique on those occasions when the weather at the destination is really down.

First of all, let's get away from an unfortunate FAA designation—*precision* and *non-precision* approaches. Sure, the word aptly defines the differences between these approaches that provide both glideslope information and/or guidance to the runway centerline, and those that merely tell you where the airport is. Unfortunately, the word "nonprecision" seems to suggest to some pilots that they can get away with sloppier flying on the nonprecision approach, when actually just the opposite is the case. The less help you get from the instruments and ground navaids, the less room there is for any error on your part.

We're not going to discuss how to shoot approaches here; ostensibly, those of you who are instrument-rated already know how to do that. The area we're going to address is those few seconds between the time we break out on the final to the time when we put a wheel on the runway: in other words, the *transition* from the approach to the visual execution of a landing.

The key for a successful landing at or near minimum is, in a sense, to delay the transition as long as possible. Let's say we break out at 300 feet, which (on a normal ILS) would put us just under a mile from the touchdown point. Look outside—there's the runway. Great. Forget about it; go back on the dials and continue for the moment as though all was gray outside.

Why? Because when you suddenly see the runway, or, more often, the "rabbit" of the strobe in the approach lights, you really don't have any reference around the airport, so it's virtually impossible to gauge your height or rate of descent with any degree of accuracy. This problem can be compounded, especially at night, by the optical effects of a curved windshield, and it gets even worse in rain.

If there's a VASI installation, of course, you can use it for guidance, but you still shouldn't forsake the instruments. A sudden patch of ground fog or scud can "turn the airport off" just as quickly as it appeared in the first place. If you have been descending visually, you might now find yourself way below the glideslope and without enough time to get reestablished before having to miss the approach.

What about when there's no glideslope? Well, that makes it even harder. On a straight-in approach, you should have a good idea of the rate of descent required to make the runway, and some nonprecision approach plates show a so-called *Visual Descent Point* (VDP) to make matters easier. What we're trying to do here is strike a happy medium. On the one hand, getting too low, too far out leads to dangerous "drag it in" approaches among who knows what obstacles short of the runway, while on the other hand, staying too high, too close leads to big rates of descent from which it's difficult to recover close to the ground. The ability to judge distances, which comes with practice and a bit of simple mental arithemtic, help a lot here. A "standard" descent path is 3 degrees, or 330 ft/nm. Thus, if you know how far you are from the airport on a visual approach, and how much altitude you have to lose, you can arrive at a descent rate to set up and stick with until you're in so close you can complete the landing visually.

Let's backtrack for a moment, and examine the approach itself. Virtually any approach, be it ILS, VOR, VOR/DME, RNAV, NDB, or whatever can be flown in such a way as to utilize a stabilized rate of descent. That stability is the key to a smooth transistion. If your rate of descent is set up properly, you should be able to hold it (using the vertical speed indicator if no glideslope is present) almost all the way to the threshold. Moreover, if you planned and set up the approach properly, you should not have any major speed reduction to contend with as the field comes in sight. The whole idea can be summed up as "don't just do something, sit there!"

ILS Below Minimums

Let's close with a short discussion of something you'll probably never be faced with—an ILS below minimums. By this I mean the situation that could conceivably arise from causes such as mechanical failure or weather that rapidly becomes much worse than forecast at both primary and alternate destinations. This may also include something like a medical emergency onboard that requires an immediate landing at an airport where the conditions could be anywhere from right on the minimums to completely zero-zero. Obviously, this isn't something that's recommended, but the time may come when you have to do it, and a little forethought, and possibly some practice when the weather's not so bad, could make all the difference in the world.

The characteristics of ILS installations differ from airport to airport, and we all have our favorites. Generally, the signals get rougher and rougher as you approach the narrow end of the approach. One reason the standard ILS minimum is still 200 feet is that, at most installations, the glideslope can't be depended on much below that decision height. An exception is those large airports that offer Category II (or, better yet, Category III) ILS approaches, since these are predicted on decision heights of 100 feet (CAT II) or less (CAT III).

Faced with the need to get down below minimums, even on a rough ILS, what can you do? Well, if there are several airports in the area in which you're going to have to descend, and one has CAT II or III capability (but all are below minimums), that's obviously the one to pick. The San Francisco Bay area is typical of this sort of choice.

Next, time and fuel permitting, take a practice shot or two at the approach, all the way down to Decision Height, then execute the missed-approach procedure. This will give you a chance to become familiar with any kinks that might exist in the *beam* (some of the best installations may have them!) as well as getting a good *handle* on the required drift and rate-of-descent corrections for your actual wind and weather conditions.

Once at or inside the outer marker, shoot these practice approaches (or the actual one, if you don't have the opportunity for practice) at a somewhat lower airspeed than you'd normally use. Remember that you may be landing at the end of the approach without ever having the opportunity to see the runway environment for a good flare. Be careful because you'll literally be "flying into the ground." Obviously, it behooves you to do this as slowly, and in as nose-up an attitude as possible. Absolute minimum speed, however, may not be the ideal way to do it. Not only does this reduce your stall margin to uncomfortable levels and make the effects of wind on the approach more severe, but, you may find, once you know your airplane, that you'll get a better nose-up attitude by using less flap, even if this requires a slightly higher speed.

Let's face it, this is an emergency procedure in any case, and any runway long enough to have an ILS is likely to be more than long enough for our class of aircraft (and nice and wide, too). Even if we go off the far end, we're likely to do it slowly!

Zero-Zero Approach

Very well, it's time for the actual zero-zero approach. Don't get too up-tight, you're not doing anything you haven't done before, you'll just be using a different set of references for the last few hundred feet. At the outer marker, have all your ducks in a row. Make sure that your speed gear and flaps are down, your prelanding check is complete, and that its properly established on course and glideslope.

From here on down, your job is to fly the approach as smoothly and accurately as you know how; keep those needles centered, and make your corrections small and smooth.

The localizer should be pretty good all the way to the ground, but on most ILS's the glideslope will start getting nervous at about the middle marker. At this point, you'd normally be able to see the approach lights. This time, though, don't divert much attention looking for them. Instead, concentrate on *freezing* the vertical speed indicator at the value it was showing when you were tracking the glideslope. Don't worry too much about the glideslope; inside the middle marker, vertical speed and the localizer/DG combination are your prime references.

An invaluable aid for this sort of thing, incidentally, is a *radar altimeter* (it's required for most CAT II and CAT III operations.) If you have a good one, you can even use it to initiate a slight flare over the runway, especially if it drives one of those nifty "rising runway" displays in your flight director.

Without it, just freeze the rate of descent and press on until you feel a bump, then you're on the runway! If you *still* can't see anything (sometimes this can be the case), keeping tracking the localizer as you carefully brake to a stop; then creep along the centerline until you see a yellow line leading off the runway, and follow it onto a taxiway. You may still be lost on the airport, but you can work that detail out with Ground Control.

Admittedly, this all may sound quite desperate, and in a sense, it is. A properly-planned flight should never end up in such straits. It's a good skill to have, though, and one easily acquired. The next time you're doing some proficiency flying in your aircraft, try taking a couple of approaches all the way onto the ground (under the hood, or course, and with a pilot qualified in your aircraft in the other seat). You'll be surprised at how easy it is!

9

Pressurization, Oxygen And High Altitude Physiology

One of the most significant factors in making the modern advanced twins and turboprops as useful, flexible, and popular as they are has been the steadily increasing availability of pressurized cabins. After all, aircraft with a certain amount of "over the weather" capability have been available since before World War II, but few passengers relished the idea of spending a few hours strapped into an uncomfortable and confining oxygen mask. Pressurization changed all that, of course, with its ability to maintain a lower-altitude cabin environment while the aircraft itself flies high in the thin, cold air. Of course, even pressurized aircraft must carry some sort of supplemental oxygen equipment to deal with pressurization failures, so we'll touch on oxygen systems in this chapter. This information is also relevant to aircraft such as the Navajo (or other unpressurized but turbocharged cabin class twins), which are usually operated at lower altitudes, but which occasionally are flown at altitudes requiring that pilot and passengers use oxygen.

The Human System

A good place to start is not with a discussion of aircraft systems, but with a discussion of a rather more basic system: yours (and mine)—the one by which oxygen is extracted from the air and made available in our bloodstream.

This takes place in the lungs (no surprise, eh?) in tiny air sacs called *alveoli* (Latin for "tiny air sacs"). The outer surface of each sac is surrounded by tiny capillaries, and the wall between them is porous, so oxygen can diffuse through it into your blood.

The basic gas laws of physics dictate that gases diffuse from regions of higher pressure to regions of lower pressure. Thus, for oxygen to get into your bloodstream, the oxygen in the alveoli must be at a higher pressure than that in your blood. The less the pressure difference, the less oxygen will get into your blood (and hence to your brain and other useful organs). Thus, at higher altitudes, you'll start to show the symptoms of *hypoxia,* or oxygen insufficiency (Table 9-1).

Table 9-1. The "Time of Useful Consciousness" for Various Altitudes.

Altitude (ft)	O_2 Rapidly Disconnected, Sitting Quietly	O_2 Rapidly Disconnected, Moderate Activity	Rapid Decompression, Sitting Quietly
22,000	10 min	5 min	–
25,000	3 min	2 min	2 min
28,000	1½ min	1 min	1 min
30,000	1¼ min	¾ min	¾ min
35,000	¾ min	½ min	½ min
40,000	30 sec	18 sec	23 sec
65,000	12 sec	12 sec	12 sec

Some people think that this is because the air at high altitudes contains less oxygen than at lower ones. This isn't true; the ratio to the other components of the air (nitrogen, carbon dioxide, etc.) is virtually the same at any altitude: about 20/80. The total air pressure at any altitude is the sum of several *partial pressures*, one for each component of the air. The problem is that at higher altitudes, there's less air altogether; while the *ratio* of partial pressures remains constant, each partial pressure, including that of oxygen, decreases (Table 9-2).

Maintaining the Proper Environment

To maintain a sea-level percentage of oxygen in the blood, then, we can resort to various technical subterfuges, all designed to maintain an equivalent *partial oxygen pressure* around 160 mm Hg (the same as at sea level) in the lungs.

One method, of course, is pressurization. This involves simply providing air at or near sea-level pressure. It would be nice if we could just have an air compressor feeding individual masks, like oxygen masks, but unfortunately these would inflate the unfortunate users like balloons as outside pressure decreased with increasing altitude. Instead, we have to

Table 9-2. The Atmospheric Pressure and Oxygen Partial Pressure as a Function of Altitude.

O_2 Partial Pressure (mm Hg)	Altitude (ft)	Atmospheric Pressure (mm Hg)
160	Sea level	760
126	5,000	632
105	10,000	523
85	15,000	429
70	20,000	349
56	25,000	282
45	30,000	225
36	35,000	179
29	40,000	141
23	45,000	111
18	50,000	87

pressurize the entire environment, as in a pressurized cabin. (Or, for that matter, a pressure suit, which is nothing more than a very small pressurized cabin that can flex with its wearer's movements.)

The other method, and one that requires no additional pressure, is to increase the partial pressure of oxygen relative to the rest of the air, that is, to provide a synthetic atmosphere with more oxygen in it than the real thing. That's what aircraft oxygen systems do. They provide a breathing mixture in which the *partial pressure of oxygen* is maintained at or near 160 mm Hg while the overall pressure decreases. The higher we go, the more oxygen the breathing mixture must contain to maintain the correct partial pressure. At 15,000 feet, for example, the amount of oxygen provided has to be about 40%, or twice the oxygen content at sea level. At about 35,000 feet, the *total* pressure has dropped to near the 160 mm Hg mark, so the breathing "mixture" has to be 100% oxygen to maintain the proper oxygen level in the bloodstream.

Two Types of Oxygen Equipment

There are two common types of oxygen equipment provided in civil aircraft: constant-flow and demand units.

Constant-flow types are most common in unpressurized twins, and they can also be found as passenger (and sometimes crew) emergency masks in pressurized reciprocating and turboprop twins. In these systems, a steady metered flow is provided to simple masks consisting of a facepiece and an expandable plastic bag (Fig. 9-1).

In the simplest masks, this bag is used as a *rebreather*. As the wearer exhales, the first part of each exhalation fills the bag, while the rest is simply discharged into the ambient air. The assumption is that the first part of each exhalation is composed of air that was in the "dead spaces" of the wearer's respiratory system, such as the mouth, throat, and bronchial tubes. Since this air never found its way into the lungs, it's still rich in oxygen. On the subsequent inhalation, this oxygen-rich air, further enriched by the additional oxygen flowing from the regulator, is inhaled first as the bag deflates, thus going all the way into the lungs, while the "dead spaces" are topped off by ambient cabin air.

More recent and/or sophisticated constant-flow masks use a slightly different principle and incorporate a "sequential valve" between the oxygen bag and mask facepiece. In these, oxygen from the regulator fills the bag only. The first part of each inhalation collapses the bag and takes this (virtually pure) oxygen deep into the lungs, while the remaining spaces are filled with cabin air. On exhalation, *all* exhaled air is vented into the cabin, since the reverse flow closes the sequential valves; meanwhile, the bag has been filling with oxygen for the next breath. Sequential masks tend to waste less oxygen while delivering more to the user, as well as reducing the "instant replay" with each breath of that garlic-and-limburger sandwich you may have had for lunch! Don't laugh too loud at this factor, there are few things more unpleasant than being airsick into an oxygen mask.

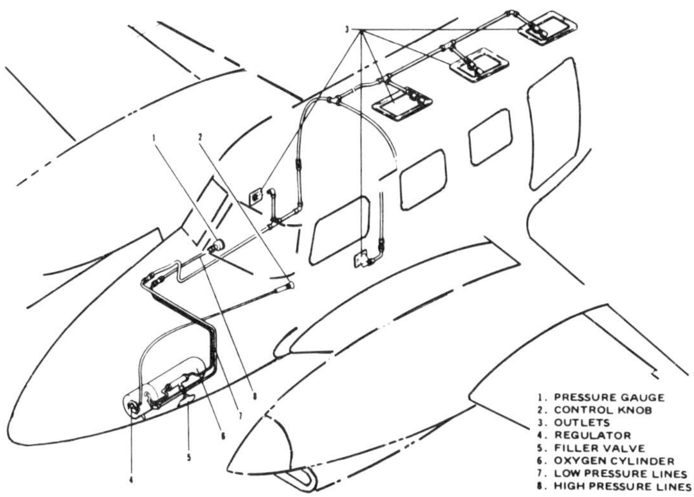

Fig. 9-1. A typical oxygen system includes an oxygen storage bottle and regulator and low-pressure lines to the various oxygen mask locations throughout the airplane. Some larger aircraft incorporate airline-style "dropout" automatic masks. Checking condition and storage of oxygen masks is an important preflight item in pressurized aircraft.

The oxygen regulators of constant-flow systems are either manual or automatic types. In the simpler manual systems, oxygen is either provided at a fixed rate appropriate for a given range of altitudes (but insufficient at higher altitudes, and terribly wasteful at lower ones,) or it can be manually adjusted by the pilot to correspond with the flight altitude. (Incidentally, the manual-adjustment types offer you a way to "cheat" a bit if you *have* to stretch your oxygen supply, since their altitude settings are predicated on maintaining near-sea-level oxygen levels in your blood. Thus, if you're running short on oxygen and still have a couple of hours to fly at, say, 20,000 feet, you can set the regulator for *15,000* feet, and your body will only get as much oxygen as it would at 5,000 feet MSL rather than at sea level. This will extend your oxygen supply. It's not a good idea for regular practice, though.)

Better regulators are adjusted automatically be an aneroid device to provide the correct oxygen flow rate at any altitude. In either case, oxygen under moderate pressure is provided to plug-in connectors around the cabin; small orifices in the individual mask connectors perform the final flow reduction. In some installations, different color-coded connectors provide varying mask flow rates. Usually, the pilot's mask has a higher flow rate and is marked with a red band.

The plastic oxygen hoses for each mask contain a flow indicator (a small device which shows a red band if flow is insufficient) to warn the user in case not enough oxygen is being delivered.

You should bear in mind that oxygen will flow from a constant-flow mask anytime it's plugged in with the system turned on. Thus, masks should only be plugged in when actually in use. The drop-out emergency masks found in some pressurized aircraft are of the constant-flow type. Generally, each is connected by a lanyard to its own cutoff valve, so that pulling the mask down to a wearable position turns on its oxygen flow, therefore, unused masks don't waste oxygen.

A more sophisticated system is the demand type. This type is usually found on the crew masks of pressurized turboprops 'and a frew recips), as well as in specialized high altitude aircraft such as photography ships and sailplanes.

In this system, oxygen flows to the mask only on demand, that is, when the wearer inhales. There's no rebreather bag, and the mask facepiece is a bit more complex, with one-way valves for both inhalation and exhalation.

The oxygen regulator unit automatically adjusts the percentage of oxygen in the breathing mixture to provide the proper partial pressure of oxygen for any altitude.

Most demand masks or regulators offer an additional capability—they can be switched to provide 100% oxygen at any altitude (or, of course, they'll do so on their own at about 35,000 feet). This can be invaluable during emergencies involving smoke or other contamination of cabin or pressurization air, or for first-aid or therapeutic purposes. (I can recall one corporate client who apparently was well aware of this fact. He'd occasionally come aboard the King Air 100 after a wild night, looking like he needed Last Rites more than first aid, and sit in the forward-most seat so that the hose of a crew mask was long enough to reach him. A few minutes of 100% oxygen seemed to make a new man of him!)

A complete discussion of the symptoms of hypoxia goes beyond the scope of this book. What's important to remember is that such symptoms can be *insidious*, you may not notice them until too late, and they can *differ from person to person*. Some pilots feel the first twinge as a tingling in their fingertips; others become euphoric, which can be particularly dangerous.

Any pilot who's planning to command an airplane, either pressurized or not that will be operated at high altitudes should attend one of the high-altitude chamber indoctrination courses offered at numerous military bases around the country and at the FAA's Oklahoma City facility. Contact the nearest GADO or Regional Flight Surgeon for details. The course usually takes a day and a half, and is time well spent. I'd like to see it required for the pilots of pressurized aircraft; perhaps you could get an endorsement on your medical certificate to that effect.

Pressurization Systems

Regardless of aircraft type, there are only a few basic elements involved in a pressurization system—a reasonably "airtight" cabin, a

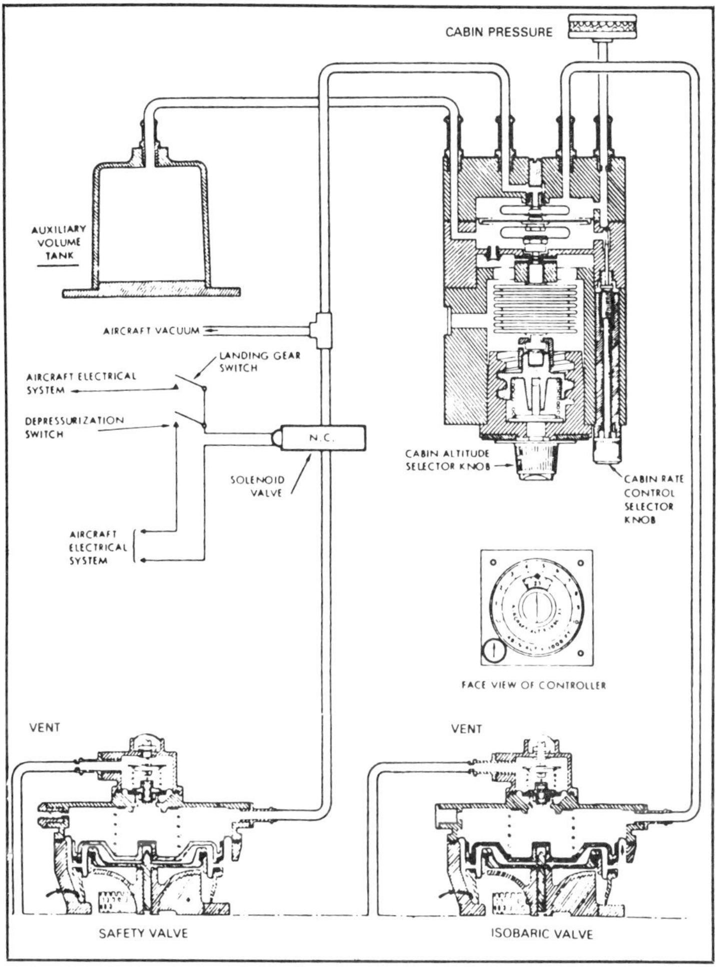

Fig. 9-2. A typical pressurization system includes a controller and two outflow valves: the isobaric valve, which regulates cabin altitude as set on the controller, and the safety valve, used to dump pressure in emergencies and/or when the aircraft is on the ground. (Note the connection to a landing gear switch, which opens the safety valve when the aircraft's weight is on the wheels.) The auxiliary volume tank reduces system surges and "ear bumps."

source of higher-than-ambient air to shove into it, and some means of controlling its pressure. (Fig. 9-2).

Of course, there's more to building an airtight cabin than just sealing all the nooks and crannies. Even a mild degree of pressurization can build up very large stresses and the cabin structure has to reflect this. Most cabin twins and turboprops pressurize to a *differential pressure*, that is, the

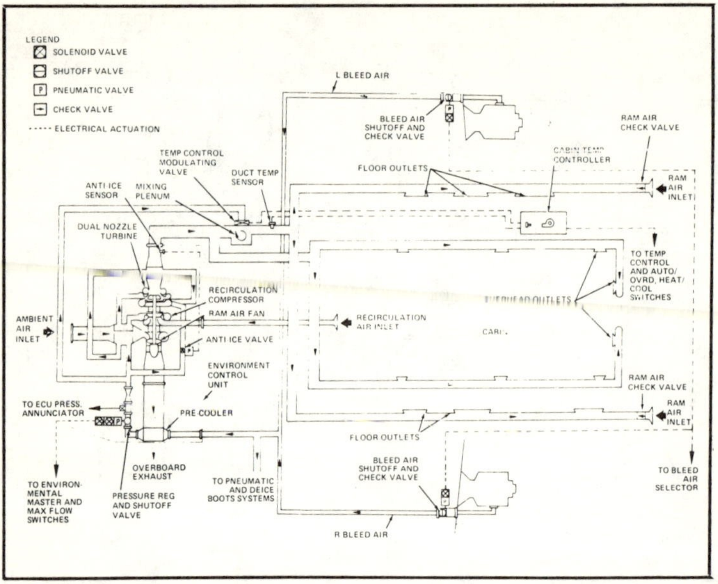

Fig. 9-3. The environmental system of the Rockwell Commander 840 uses an air-cycle machine, or dual-nozzle turbine, to provide cold bleed air for ventilation and pressurization. Warm bleed air is mixed with it to reach the desired cabin temperature. Some larger aircraft, like the Merlin III, have a separate air-cycle machine for each engine's bleed air system.

difference between cabin pressure and the Great Outdoors, of around 5 psi. At that pressure, a foot-square window (or, for that matter, a square foot of cabin skin) is subjected to a force of some 720 lbs trying to blow it out of the airplane! No wonder, then, that structures have to be rugged. The doors and windows of these aircraft tend to be smaller than those of unpressurized airplanes, and they have rounded corners to reduce the buildup of strain in critical areas.

Door latching and sealing mechanisms also have to be pretty sophisticated. For example, the King Air C90 has an entrance door measuring 27″ × 51½″. At the normal 4.7 psi differential pressure, the latches holding that door shut have to contend with a force of 6535 lbs, that's over 3¼ tons (or more than the gross weight of a light twin)!

That's why the doors of pressurized aircraft are more complex. The latch pins on these doors usually come out all the way around the door rather than just across from the hinge. In fact, once the door's closed on many airplanes, the hinge might as well not be there. The only reason it's really there is to keep the door from falling off when it's opened.

With a door like this come specific procedures for opening, closing, and checking it, and you should be aware of the ones for your particular airplane. Often, small inspection windows are provided through which you can check proper engagement of latching pins and safety devices before pressurizing the cabin. Door *seals* are often inflatable pneumatic units powered by pressurization bleed air or other aircraft sources, and while some are automatic, others must be inflated by the pilot before pressurizing the airplane.

Very well, now that we have our airtight pressure vessel, how are we going to pump it up?

With bleed air from the powerplants, that's how (Fig. 9-3). If we're flying a recip, its turbochargers are sufficiently "oversized" to provide air for the cabin, as well as the engines themselves. Turboprops use air bled from the engine compressor section. Again, this bleed is included when the airframe/engine combination is designed.

Some of the oldest pressurized recip twins, such as the "AltiCruiser" Commander 720 or certain versions of the Grand Commander, used mechanical, rather than turbo, superchargers. These could not provide sufficient airflow for cabin as well as engines. Instead, hydraulic pumps on the engines drove a hydraulically-powered cabin supercharger in the aft fuselage.

Environmental Control System

At any rate, we've now arranged for some source of high-pressure air. As it comes from the engine or supercharger, it'll not only be at a much higher pressure than we want in the cabin, it'll be pretty hot, too. So, in addition to controlling cabin pressure, we'll need some sort of environmental control system to maintain cabin temperature at comfortable levels (Fig. 9-4).

Let's deal with pressure controls first. It's impractical to control how much air is pumped into the cabin from the power plants, since changing power settings would cause the cabin altitude to change, and also, changes in cabin pressure requirements would, in turn, affect the powerplants. Instead, pressurization control systems allow all available bleed air to flow into the cabin, then they control the rate at which it leaks out. This, of course, also provides a constant "throughput" of air for proper cabin ventilation. Otherwise, it would get very stuffy by the end of a long flight.

The actual controlling is done by *outflow valves* mounted somewhere on the pressure vessel, usually a forward or aft bulkhead. Most aircraft have two of these valves. One provides the actual control, while the other one (often a simpler type with only "open" and "closed" positions rather than the variable feature on the actual control valves) acts as a safety valve to prevent overpressurization. The safety valve can also be opened by the pilot for cabin depressurization, and is often actuated automatically by a *squat* switch on the landing gear to ensure that the cabin isn't pressurized with the airplane on the ground. Know where the outflow valves are in your airplane, and be sure nothing is placed where it can block them. One pilot of my acquaintance had to depressurized in a hurry, to the detriment of his and his passengers' ears, when one of those plastic hanging clothes bags was sucked over both outflow ports in the light pressurized twin he was flying.

Let's examine the pilot's pressurization controls for a typical aircraft. It might be either recip or turboprop, but the basic system will be the same.

Each pressurization source has an individual shutoff valve to allow it to be secured in the event that it starts providing contaminated air. (If it does, by the way, it's a sign that all is not well with that particular powerplant and you may want to shut it down. In a recip, contaminated pressurization air can mean an incipient engine fire and/or a failure of an induction duct or filter; or it can mean that an oil seal or bearing is breaking down in the turbocharger itself. Contaminated air coming from the bleed of a turboprop powerplant could mean impending bearing failure or engine fire.) In some aircraft, the bleed shutoff is a mechanical handle, while in others it's electric, either with individual switches or a left-both-right-off selector. When a pressure source is shut down, its air is often dumped overboard, which can lead to some confusion. Some source-shutoff valves are labelled "on" and "dump," but activating them to the "dump" position will not necessarily dump (i.e. depressurize) the cabin.

A control, on the other hand, that will dump the cabin pressure is the aptly-named "dump valve." This mechanical valve is also sometimes called "ram-air" valve, in that it does double duty. Not only is it a means of depressurizing the cabin, but since it's usually installed in some sort of forward-facing scoop, it can provide ram-air ventilation for unpressurized operation when both bleed air sources are shut down. Not all airplanes have manual dump valves. Often, those that do are ones in which a total electric failure causes the cabin to pressurize (rather than depressurize), so the dump valve serves as a means of depressurization even if the electrical valve is out of commission.

The normal control for pressurizing and depressurizing is one of the two outflow valves, and its function is controlled electrically by a *pressurize-depressurize* siwtch on the instrument panel. The same circuitry is normally hooked to a landing-gear weight switch to depressurize the cabin automatically whenever the airplane is on the ground.

The actual cabin pressure controller is a pneumatic or elctro-pneumatic device hooked to the main cabin outflow valve. In some of the very simplest airplanes, this device isn't accessible to the pilot at all; instead, it's preset for some specific cabin altitude. The airplane operates in unpressurized mode until that altitude is reached. At that point, the cabin "levels off" while the airplane continues to climb. As the altitude is reached at which the cabin is "pumped up" to maximum differential pressure it begins to climb once again, although at a slower rate than the airplane.

Most current control systems, however, are more sophisticated, and offer variable altitude selection. The simpler ones have no separate rate control, and their operation is virtually instantaneous so a certain amount

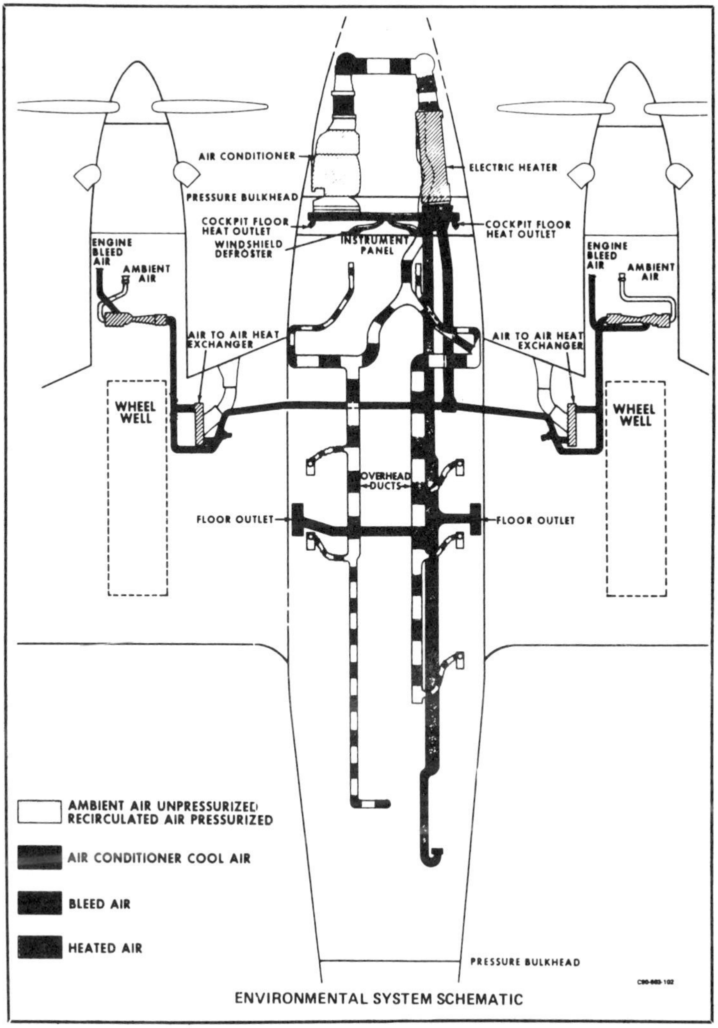

Fig. 9-4. In the King Air environmental system, warm engine bleed air is supplied for all pressurization, ventilation, and heating. An electrically-powered Freon-cycle air conditioning unit provides cooling as required.

of caution is required to prevent "ear bumps" (we'll go into operational techniques in a moment). The better ones have rate controls, by which the pilot can adjust the rate at which the cabin altitude climbs and descends.

With either type of controller, the pilot can set the desired cabin altitude, and the system will maintain that altitude within its capabilities. Of course, it cannot pressurize the cabin to more than the allowable maximum, so there comes a point at which the cabin will climb to an altitude above that set on the controller. Moreover, no system can maintain the cabin at a higher altitude than that of the airplane itself.

Most controllers have a handy additional scale which reads aircraft altitude as maximum differential pressure for any give cabin altitude setting. As we'll see in a moment, this simplifies setting the pressurization system for optimal performance.

Operating Pressurization Systems

Properly managed, a pressurization system can contribute immeasurably to comfort while minimizing additional pilot workload. Improperly managed, it can not only cause lots of hassle, but it can leave you and your passengers with your ears ringing or blocked. Like the airplane itself, the pressurization system is easiest to handle if you can stay ahead of it at all times.

We'll more or less ignore the old fixed-altitude systems since there's not much you can do with them anyway. All you can do is just fly the airplane, and the cabin will hold the present altitude as long as it can. The variable systems allow us a certain amount of flexibility. In a sense, the simpler systems, those without a rate control, require more pilot effort and finesse to operate. Choice of the proper altitude setting is important.

One very simple way of operating a fixed rate system is to choose a single setting for the entire flight. This setting can be the altitude of the landing airport plus, say, 500 to 1000 feet, or it can be the lowest cabin altitude that the airplane can maintain at the planned cruise flight level, whichever is *higher*.

Operated in such a fashion, the system is more or less a "set and forget" one. All you have to do is make sure the main switch is set to *pressurize* before takeoff. On a flight, for example, from Los Angeles (sea level) to Denver (5300 feet), you might set the dial for 6000 feet cabin altitude. The cabin would climb with the airplane to 6000 feet beyond which the system would act like a fixed setting type, depressurizing on approach some 700 feet off the ground.

What about the return flight, though? Well, of course, the simplest way would be to leave well enough alone, but you'd lose the benefits of pressurization once you were below 6000 feet on the approach into Los Angeles.

Moving the pressurization control of a fixed-rate system inflight will result in almost instantaneous changes in cabin altitude, so it's generally a bad idea unless your passengers like having their ears climb and descent at 30,000 fpm or so. There is, however, one exception: if the system is already operating at maximum differential, it can be set for lower altitudes with no immediate effect. Let's say, on our flight from DEN to LAX, that the system (when set for 6000 feet) reaches max differental pressure at FL 190. Once we've climbed above that level, we can set the controller for, say 2500 feet, and nothing will change. As we descent below FL 190, the cabin will start down from 600 feet for 2500 at a rate proportional to, but less than, the aircraft descent rate. Similarly,the system can be set for *higher* cabin altitudes enroute as long as the new setting doesn't bring the system below maximum differential. For example, we could depart Los Angeles with 2500 feet set into the controller, climb to FL 190 (at which altitude a 6000-foot cabin requires max differential pressure), and then reposition the dial at 600 feet for the later landing at Denver without causing ear bumps.

Variable-rate controllers can be set inflight at any time, although, of course, they're still subject to the same limitations of maximum differential pressure. Normal practice is to set them before takeoff for the lowest altitude the cabin can maintain at the planned cruise altitude, adjusting the rate control so that cabin and airplane "level off" at about the same time. Once cruise altitude is reached and the system's running at max diff, you can reset the cabin altitude selector for a lower (landing) altitude, adjusting the rate control on descent to keep the cabin "ahead" of the airplane.

Sound confusing? Here's an example: we're flying a King Air, with 4.7 psi max differential capability, from LAX to DEN. We're going to cruise at FL 230, at which altitude we can maintain about 8500 feet cabin altitude, so that's what we'll set before takeoff. For purposes of discussion, let's say it'll take us 15 minutes to climb to cruise altitude. The cabin has to climb 8500 feet in those 15 minutes, so after takeoff we'll check the cabin rate-of-climb and, if necessary, reajust it for a climb rate of about 560 fpm; 600 fpm will be close enough. If the cabin "beats" the airplane to cruise altitude it'll just level off and wait for the airplane to catch up whereas if the opposite occurs, the max differential presure will be reached before level-off and cabin rate of climb will increase for the last part of the climb.

Enroute, we have plenty of time to figure out that the cabin will have to descend from 8500 feet to 6000 feet on approach. A normal descent profile willl usually start the airplane down a good distance out so maximum differential pressure is seldom a problem. In other words, the airplane can usually start down before the cabin. Three hundred fpm is a good descent rate for ear comfort, so we have 2500 feet to lose, so we consult our trusty whiz wheel once again and start the cabin down when we're about 8½ minutes out on the approach, perhaps a bit more just for "slop." The caluulations aren't at all critical, and slight rate changes can be made without causing anyone discomfort. What we want to guard against is getting the airplane down ahead of the cabin, since as soon as that happens, cabin descent rate will catch up to the actual, and often rapid, aircraft descent, and descents are much harder on ears than climbs.

There's an additional consideration to bear in mind when flying a recip—while there's sufficient bleed air from most turbines to maintain full pressurization even at idle, recips have to be kept somewhat "spooled up" for proper operation. Most systems are full upper-deck pressure for the pressurization system, so a good rule of thumb is not to let the manifold pressure go more than five inches below a figure gained by subtracting the desired cabin altitude (in thousands of feet) from 30. Thus, for example, if you want to maintain a 3000-foot cabin, you'd plan to keep at least 22 in Hg power on during the approach.

From this, you can see that it sometimes becomes something of a chore to keep the cabin pressurized to a low altitude on those days when ATC wants you to "chop, drop, and stop." Depending on your aircraft and its operating characteristics, 4000 feet or so may be a reasonable "lowest practical cabin for most operations."

Pressurization System Emergencies

When a pressurization system fails to do its job, it does so in one of two ways: too much, or not enough (special case: not at all!). Each type of failure has its own dire consequences, and each has its own recovery techniques.

Let's look first at the "too much" mode, or over-pressure situation. This may occur gradually, in which case you may only notice it from a look at the differential pressure gauge (usually combined with the cabin-altitude gauge, although it's a separate instrument on some airplanes), or it may occur suddenly, in which case your ears will let you know right away!

The big danger in an overpressurization, of course, is structural failure. This could result in a "blowout," as it were, of a door or window, or even of primary aircraft structure. What do you do if you suspect or verify an overpressure?

First of all, check the differential pressure gauge. If if it's below redline (but the cabin altitude is lower than the one you've set), sit tight for a moment and watch to see if the pressure stops increasing before reaching redline. If this is the case, you're dealing with a failure in the controller or the primary outflow valve, and there's no reason to depressurize. The cabin will stabilize at maximum differential, and you can continue to the destination, monitoring pressure closely.

If it *does* go over redline, however, you're going to have to reduce it right away. Each aircraft's emergency procedures list may have a different idea. I tend to deal with an overpressure in stages.

First, it may be that the outflow valves have failed while closed, but not quite all the way. In such case, they may still be open far enough to handle the airflow from one bleed source, if not both, so turn one off (either one is fine). If the pressure stabilizes at some less-than redline value, you may elect to continue at altitude, but watch that pressure gauge like a hawk!

If, on the other hand, the pressure continues to rise, you have no recourse but to depressurize. In doing so, you have, in effect, changed the "too much" situation into a "not enough" one, and you can proceed from there.

Very well, what do we do if there's not enough pressure? If the difference is only slight, if we've cancelled for 8000 feet and the system is giving us 8700, it's more a maintenance hassle than an operational problem, and frankly, such occurrences are far from uncommon. Leaking door seals seem to be the most usual culprit.

What, though, if you're flying along and the cabin suddenly starts to climb? Well, there are a number of things you can check, but there's always one thing to do first: *get on oxygen at once*! Don't worry about the passengers at the moment; they might pass out, but you'll do them a lot more good if you remain conscious for the moment. Anyway, they could probably use a nap, and it'll keep them out of your hair wile you deal with the problem. (Actually, of course, they, too should get oxygen as soon as possible, but your main job is to fly the airplane.)

Then, if the cabin climb is gradual, get the passengers on oxygen and investigate the cause. Did you bump the doorseal switch with your knee, or, for that matter, the controller? Are there any unusual hissing noises that suggest a leak somewhere in the cabin or system plumbing? Are both air bleeds on?

On the other hand, if the cabin climb is rapid, or you've experienced an explosive decompression, get the airplane down to lower altitude at once. Each manufacturer, again, has their own emergency descent. Most descents involve some variation of the "get the power off and the nose down, and let the airspeed built up" theme. Here's one area where you can go ahead and chop the power in a recip, since you're not getting any pressurization anyway! I generally teach my students to start a gently roll-off to the left a sort to tame split-s, for an emergency descent. Not only does this let you see where you're going (assuming V_{mc}, of course), but it keeps the airplane under positive g to prevent scattering possible unsecured passengers around the cabin.

Actually, it's highly unlikely that airplanes of the class we've been discussing would experience a real, live explosive decompression. That's the sort of thing that happens when you accidentally jettison the canopy of your trusty F-104 at 50,000 feet. What could happen would be an extremely rapid decompression, say, if a window were to blow out. The cabin would run up to actual flight altitude in a couple of seconds at most.

What's it like to ride through a rapid decompression? It's not all that bad, as you, too, will find out when (not "if"—*when*) you take the altitude chamber course. There's a hell of a bang, of course, and a rush of air. Your ears will pop, and you may feel other trapped body gases expanding and leaving you through a couple of orifices. It will suddenly get very cold, and if the air is at all humid, a thick fog will form almost instantly, and will dissipate just as fast. Loose objects, clothing, Jepp binders, and so forth will get blown around.

What's important to remember here is that rapid decompression *greatly reduces your time of useful consciousness* compared to gradual depressurization. It may only be a couple of seconds, *so get the oxygen mask on before you do anything else*. From then on, you can proceed as in any other depressurization incident, although before getting up to a high airspeed on an emergency descent, you'd better check to see whether it was severe structural damage that caused the rapid decompression in the first place.

Air-Source Contamination

The other problem that could crop up in a pressurized airplane is contamination of the cabin atmosphere by smoke or fumes. This could come either from some sort of fire in the cabin itself, in which case it should be dealt with by the manufacturer's recommended cabin fire procedure, or from either of the pressurization sources.

Therefore, the first thing to do is to shut both of them off. This, of course, will depressurize the cabin, but not nearly as fast as opening a dump valve would. As the bleed air into the cabin shuts off, the outflow valve will close, and it'll take quite a few seconds for the residual pressure to vent away. This will give you plenty of time to put on an oxygen mask (and, if available, *select 100% oxygen immediately*.)

If the fumes persist, the cabin can be ventilated very effectively by opening both dump and ram-air valves, and possibly a cockpit storm window as well. Once the air has been cleared, leave the pressurization switch in the "depresurize" position, and turn on one bleed source. If the air again fills with fumes, turn in off and try the other one, repeating the cabin venting procedure if necessary. If the air stays pure, don't try the other source, since that's the most likely source of trouble; one will do fine for the rest of the flight. Monitor the engine on the contaminated side carefully for any signs of fire or bearing failure.

Re-Pressurizing at Altitude

One point that's sometimes not well covered in flight manuals is a procedure for re-pressurizing an airplane in flight after it's been depressurized for some reason. This move can be a real ear-basher if done wrong. Consider, for example, that if you're flying at FL250 with the pressurization set for 8000 feet and re-pressurize, the cabin will zip down those 17,000 feet in less than a second! Here are a couple of tips on avoiding the feeling that your ears have traded places through the middle of your head.

The easiest way to repressurize is to descend to an altitude *lower* than the highest available on your pressurization controller. Merely set in a cabin altitude higher than actual airplane altitude, turn on the bleed air source(s), close the ram air and dump valves, and climb back to whatever altitude is necessary. Some rate-control systems, too, have rate controls that operate far beyond the actual altitude control range, so that you can repressurize with the rate control set at or near minmum, and the cabin will descend "back onto the dial" at some reasonable rate.

For airplanes without that kind of rate control, though, there are still a couple of wrinkles you can use, if you have mechanically-controlled valves for either ram or bleed air.

With mechanical bleed valves, the technique is to set a high altitude on the controller, switch the safety outflow valve to *pressurize*, (with bleed air OFF), close the ram-air valve (if any), and then *slowly* open *one* air source valve. By "slowly" I mean *barely* ease the valve in from its full-closed position, watching the cabin rate of climb indicator. You'll have to make tiny corrections to keep the cabin rate of descent under control, and the whole process may take several minutes, so if you have a co- or autopilot, this is a good time to take advantage of him/her/it.

With electric bleed control, but a manual ram air control, you can achieve the same effect—leave the ram dump *open*, turn on *one* air bleed, then *slowly* close the ram valve. What you're doing in either of these procedures in effect, is to act as a human rate controller.

Environmental Systems in Pressurized Aircraft

The heating and/or air conditioning systems in pressurized aircraft are very similar to those in unpressurized ones, but since the entire system must be maintained at a higher-than-ambient pressure, there are a few additional wrinkles. Moreover, many of these systems make use of the heat produced by compression in recip twins for cabin heating. In turboprops, bleed air is rather warm anyway, and often can be cooled but the coldest weather.

Let's follow the course of pressurized air from the turbocharger to the cabin in a typical recip twin. From the upper deck side of the blower, the cabin air supply is ducted through a specially-shaped venturi, or "sonic nozzle," which allows it to expand, reducing its pressure and temperature, as well as its flow rate. It's still pretty warm, though, and in most aircraft it's ducted through an intercooler or heat exchanger, in which it can dump some of its heat to the outside air. From here, it's fed to the cabin, either directly or via the heating and cooling system, depending on the aircraft.

The intercooler is a point at which air temperature can be controlled quite readily, and in recip twins this is done with doors or flaps that either direct outside air over the intercooler or shut it off. Note that pressurization air continues to flow through the inside of the intercooler, but cold outside air can either cool it from the outside, or bypass it altogether.

From here on in a recip twin, there's not that much difference between pressurized and nonpressurized types. In the simplest, pressurized air is simply ducted into the cabin; if heat or air conditioning are required, these units simply recirculate cabin air. In most twins, though, pressurized air is ducted into the cabin *through* the heater and /or air conditioner, which are merely turned on and off as necessary. Thus, the airflow from the bleed system is available for ventiltion even when the airplane is operated in a nonpressurized mode. In many of these airplanes, actual heating of theair (by the usual sort of combustion heater) is seldom required, except in the coldest conditions. Normally, it suffices to simply close off the outside air supply to the intercoolers (by cockpit control) to raise the temperature of the incoming pressurization air to very comfortable levels.

In fact, even with intercoolers, pressurization air can be uncomfortably warm in summertime. Pilots of airplanes without air conditioning sometimes will alert to delay pressurizing until, say, 8000 feet on climbout in order to achieve better cabin cooling in the unpressurized mode with outside (ram) air.

In the turboprop aricraft, the situation becomes a bit more complex because the incoming bleed air from the engines is almost invariably too hot to simply be used "as is." Some form of cooling must be provided in the circuit.

In its initial stages, turbine bleed air undergoes similar treatment to that of recips—it's tapped off the engine, ducted through a sonic venturi, and sometimes mixed right at the engine with ambient air (using a jet pump arrangement) to slightly reduce its temperature. Then it's ducted on toward the fuselage, often through intercoolers in or near the wing roots. As in the recip ships, some turbine ships make use of these intercoolers as adjustable cooling devices, usually by valves which can bypass bleed air around the intercooler for increased warmth when required. In some installations (e.g. King Air 200), the bypass valves on the intercoolers are closed, one at at time, by the automatic temperature control in the aircraft environmental system.

Once intercooled, the air proceeds to the rest of the environmental system, and here's where its further fate can differ widely depending on the aircraft. In some, it enters a heating and cooling plenum where it's mixed, if necessary, with recirculated air from the cabin that's been cooled by a Freon air conditioning system. Again, the various and sundry King Air models are one of the holdouts for this sort of system.

In other twins, though the incoming hot bleed air (or at least most of it) is ducted through a device variously called an *Environmental Control Unit*, (ECU), an *Air Cycle Machine*, or a *Bootstrap Turbine*. These are all names for the same fascinating gadget, which is an air-conditioning device that uses excess energy in the hot bleed air itself to cool it (hence the "bootstrap" name).

The ECU works on the principle that expansion = cooling. Air entering the ECU is compressed by an impeller very similar to that of a turbosupercharger increasing both its pressure and temperature. This hot, high-pressure air is now ducted through a heat exchanger, where it dumps heat to the ambient atmosphere. What we now have is cooler, but still high-pressure air. The air is now allowed to expand, thus losing both pressure and temperature, through a turbine which not only provides it with somewhere to expand, but also fulfills the useful function of driving the compressor (and often a cooling fan for the heat exchanger as well). The result of all this compressing and expanding is a stream of cold, low-pressure air, so cold, in fact, that for normal cabin temperature control, a portion of hot bleed air is ducted *around* he ECU and remixed with its output. The combined, perfectly-controlled air is then fed into the cabin distribution system.

There are few particular areas of caution for a pilot running an ECU in his or her ship. If anything, these units can require less maintenance than the Freon systems (they have only one moving part, as well as saving weight). They *do* spin at very high speeds, so maintenance people should be aware that their lubricaton can be critical. If they fail, the aircraft system has some provision to bypass the ECU, so that pressurization will continue, although the cabin may get a bit warm. It behooves you as a pilot to know whether your ship uses an air cycle machine or a Freon cycle for cooling, since comfortable ground operation requires different techniques, dpeending on the system. With a Freon unit, for maximum ground cooling, you should turn *off* the bleed sources and run ambient and/or cabin air through the system, while with an ECU, you leave the bleed air *on*. In fact, the more bleed air you have, the better and several ships have "ground only" extra bleed flow valves to maximize ground cooling.

A amusing, but embarassing, occurence can show up in ships using ECU's that have *max flow* positions that don't shut off automatically on takeoff. With a cool cabin temperature selected and the ECU in max flow a muggy, humid day, rapid advancement of the power levers can cause momentary localized rain, and even sometimes snow flurries, near the cold-air outlets in the cabin! This happens when the ECU suddenly spins up to higher speed as the bleed airflow increases. This can be avoided by selecting a "flight" position on the bleed air source selected before lining up for takeoff.

10

An Introduction To Turboprop Engines

If the turbocharged reciprocating engine has brought unparalleled efficiency and economy to the class of cabin and pressurized twins, the turboprop engine has brought similarly unparalleled levels of performance and comfort as well as hitherto unattained levels of economy for turbine powerplants, while there's a lot to say for the sophisticated recips now available, there's no doubt that turboprops end up with the prize for performance, comfort, smoothness, and excess power; power which translates into comparatively sparkling single-engine performance. Look, for example, at the Beech Super King Air 200. At a gross weight of 12,500 lbs, after losing an engine at sea level on a standard day, it will still climb at 750 fpm.

In this chapter, we'll start by looking at some of the basics of turboprop engines as a whole; then we'll get into their operating controls and indicators, as well as some of their associated and specialized propeller systems, and then we'll add some practical operational techniques.

Comparing Turboprops and Gas Turbine Engines

Ultimately, a turboprop is nothing more than a basic gas turbine engine, that is, a turbojet, arranged such that its maximum power output appears as horsepower delivered to a shaft, rather than as thrust. The principles of operation are the same as for a turbojet, though. In fact, one popular small turbofan engine, the JT-15 series which powers such jets as the Cessna Citation, has much in common with the PT-6 turboprop from the same manufacturer.

A gas turbine engine operates on a four- "stroke" cycle, very much like a reciprocating engine. Of course, there are no "strokes" as such, since all moving parts of the turbine engine merely spin, and airflow through it is continuous and in one direction. Even so, we can equate the four strokes of the piston engine with similar phases in the turbine.

In the *intake* stroke of a recip, the piston is moving downward, drawing fuel and air into the cylinder. At the comparable point in the turbine cycle, ram air is scooped up and ducted into the compressor section(s) of the turbine engine.

During the *compression* stroke in a recip, the piston is moving back upward, compressing the fuel-air mixture. At the same point in the turbine cycle, the air is passing through the compressor, and its pressure and temperature are increased.

During the *combustion* or *power* stroke, the mixture in the recip has been ignited by the spark plug and burned, and the resultant expansion is driving the piston down once again. In the turbine, the compressed air has been fed into the combustion chamber, fuel has been sprayed in at the same time, and it's been burned. No ignition is necessary, since in the turbine, it's a continuous process; once a flame has been lit in the combustion chamber, it stays there.

Finally, in the *exhaust* stroke, the piston comes up once again, driving the exhaust gasses from the cylinder. In the turbine, the exhaust passes from the combustion chamber through one or more turbine wheels, and on out the tailpipe. In a pure turbojet, this turbine extracts only enough energy from the exhaust to run the engine's compressor sections, leaving the rest as thrust. In a turboprop, the turbines are more elaborate, and extract considerably more power from the hot gas stream; enough to run not only the compressor, but also to deliver reduction gears. Some residual thrust power can still be gained from the exhaust, of course, and the tailpipe is accordingly pointed aft.

In many ways, flying a turboprop differs very little from flying a recip. The basic flying skills are the same, of course, and many of the "common sense" guidelines apply equally to either type. Power management is a bit different but certainly no more difficult; on the contrary, many turboprop powerplants are significantly simpler to operate than recips. They're certainly much easier to run than recips of similar power, as well as being much lighter for the same horsepower output. Starting, too, is easier, especially when it's cold. Although some precautions have to be observed, many modern turboprops are literally as easy to turn on as a radio, just flip a switch or push a button! Even those that require a more involved starting procedure take away that element of doubt we have with the big recips, that "will it start?" syndrome we feel with each prop blade that goes by.

Two Classes of Turboprops

Turboprop powerplants can be divided into two basic classes, with both well represented in the aircraft market. These are free-turbine types, such as the little Allison 250 (originally a helicopter powerplant, but now finding its way into more and more fixed-wing applications), the recently-introduced Lycoming LTP-101 series, or the ubiquitous Pratt & Whitney/United Aircraft of Canada PT-6. In a sense, this last engine is the archetypical small turboprop, partly because it powers the Beech King Air. This is the first lighter-than-airline turboprop of note, and one of the progenitors of the whole breed.

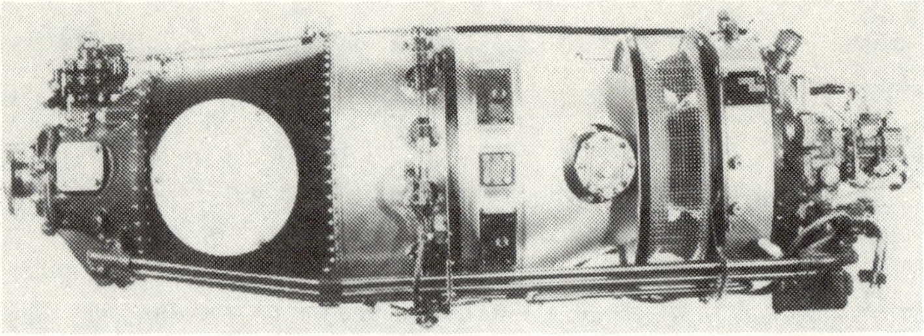

Fig. 10-1. Exterior view of a PT-6 engine; airflow is from right to left. The prop governor is mounted on the prop gearbox at left, while the fuel control is on the accessory gearbox at right. The round white cover seals off the exhaust; intake is through the grilled area left of the accessory gearbox.

Unfortunately, there's no catchword that means the opposite of *free turbine*. There's a whole class of *non-free-turbine* engines, though, with the various representatives of the Garrett TPE-331 series the most numerous, powering such ships as the Turbo Commanders, Merlin/Metro series, and the Cessna Conquest. (You can order one version of the King Air with them too!). Similar foreign engines include the Turbomeca Astazou (used on Handley-Page Jetstream) or, in a slightly larger power class, another elder statesman of the turboprop world, the Rolls-Royce

Fig. 10-2. Here's a look under a King Air cowling at the turbine end of the PT-6 engine. The paired pipes running around the engine are the fuel lines feeding the combustion chamber nozzles; the electrical line (covered with silver braid) just below center is one of the igniters, used for starting. The chain-and-sprocket arrangement operates the inlet ice separator vanes.

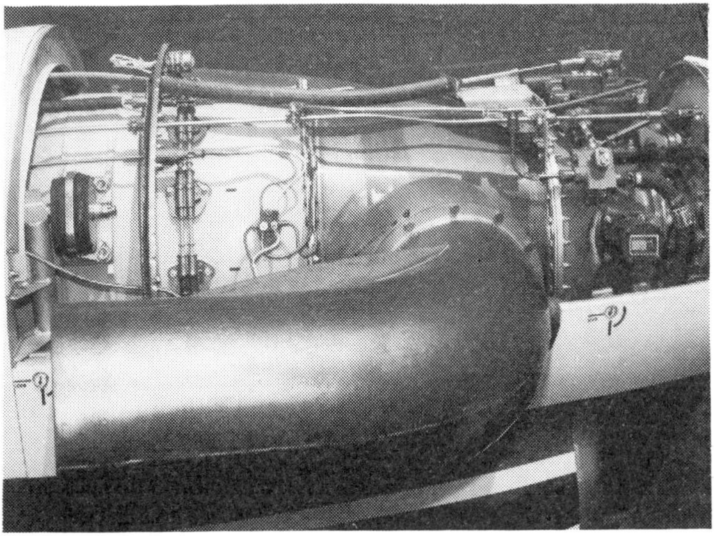

Fig. 10-3. Here's another view of the forward end of the PT-6, with the turbine just left of center. The exposed wires run the ITT indicator, while the silver device at the top is an engine fire detector. At the left is a rubber engine mount. To the right of the exhaust section, with its big outlet duct, is the propeller gearbox with prop governor and linkages.

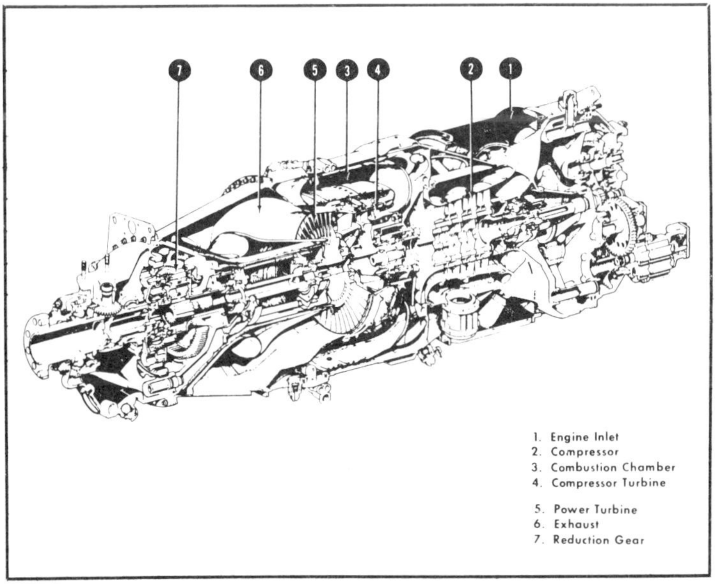

Fig. 10-4. The PT-6 is a reverse-flow, free-turbine engine. Air enters the rear of the engine and moves from the right toward the prop shaft at the left.

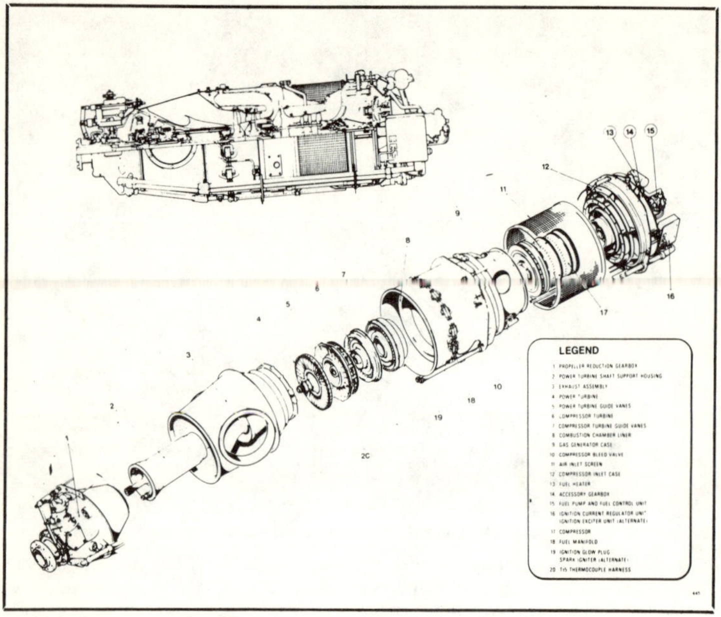

Fig. 10-5. Even a complex machine like a PT-6 engine breaks down into a relatively small number of simple subassemblies.

Dart, which powers such corporate pleasure domes as the Grumman Gulfstream I.

What's the difference between "free turbines" and *direct-coupled* engines? It's a simple but fundamental one.

Free-turbine engines, such as the PT-6, use two separate turbine systems (Figs. 10-1 through 10-7). As the air emerges from the combustion chambers, it first passes one turbine whose shaft runs back to the compressor section. In a sense, the entire engine up to this point can operate as a separate entity of its own, which we can call a *gas generator*. Even if the propeller and reduction gearing weren't there, this "sub-engine" could run on its own, generating huge flows of hot gas.

There's more to it, of course. As soon as it's passed through the first, or *gas generator*, turbine, the flow impinges on *another* turbine. This is the "free-turbine," or power turbine, and it's this one that runs the propeller, through an entirely separate shaft and set of reduction gears. There's no direct mechanical connection between the two turbines, hence the *free turbine name*.

In a direct-coupled engine, on the other hand, there's just one turbine (although in either engine, when I speak of a turbine I may mean several stages of wheels, rigidly connected on the same shaft), which runs both the engine compressor section *and*, again through appropriate reduction gears, the output shaft and propeller.

The reduction gears for either engine, by the way, represent some pretty sophisticated engineering in their own right. The trend in turboprops seems to be toward slower and slower prop rpm (for noise and efficiency considerations), with some now barely turning up 1600 rpm at full chat. At the same time, the compressor/turbine shaft (or the power turbine in a free-turbine mill) might be cranking around at better than 40,000 rmp. As we'll see a bit further on, this can lead to some rather interesting windmilling drag.

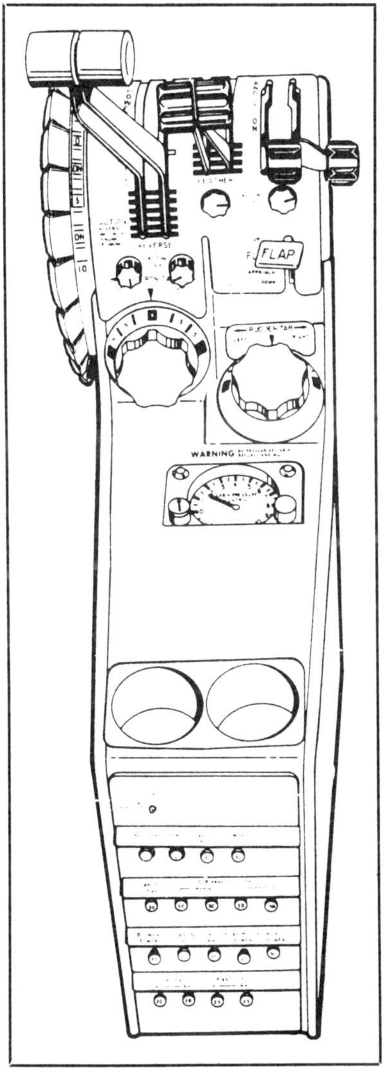

Fig. 10-6. On the power quadrant of a PT-6 powered ship, in this case, a King Air, the power levers control forward and reverse thrust; the prop pitch levers work just like those on a recip. Feathering an engine does not shut it down automatically; it'll keep running quite happily. The condition levers set engine idle speed and turn the fuel on and off for startup and shutdown.

Incidentally, you can often spot which kind of engine an aircraft has, even from fairly far away, by observing its propeller. In a freee turbine, the *core* engine can start readily and accelerate toward idle speed more or less independently of what the propeller may be doing, since there's no direct mechanical connection. Thus, many props used on free-turbine engines don't have centrifugal feather stops like those on recips. Therefore, when the engine shuts down, oil pressure drains from the propeller hubs and then feather as as they coast to a stop. See a turboprop sitting on the ramp with both props feathered, and chances are very good that it's running a free-turbine engine.

Direct-coupled engines, on the other hand, have enough trouble just accelerating the mere *mass* of a propeller in those first relatively weak moments right after lightoff. The situation would be worse if they had to drag feathered blades sidewise through the air until the oil pressure came up. Thus, the prop used with this type of engine generally has a "start lock" or anti-feathering stop, set automatically by centrifugal force during spooldown, that holds the blades in flat pitch when static or for start. Not that they can't be feathered, of course, they can if necessary, but they don't do it by themselves every time you stop an engine.

Each type has specific advantages and disadvantages. Direct-coupled engines tend to be a bit higher and physically smaller, since there's one less turbine stage to contend with. Just look at the ridiculously tiny engine nacelles of, say, a Mitsubishi MU-2, or the skinny cowling of an Astazou, on which the prop spinner is the largest-diameter part. On the other hand, they tend to be a bit noisy, especially on the ground. Since the turbine engine is only happy over a relatively small range of speeds, direct-coupled engines have to run pretty fast all the time, even for taxi, with power output controlled largely by prop pitch. They're also a bit harder to start at times, since the starter-generator unit (which we'll discuss in a moment) has to accelerate not only the engine, but the reduction gear and propeller as well for every start.

Free-turbine engines, then, are a bit larger and heavier, and sometimes, depending on size and manufacturer, they cost a bit more. On the other hand, they're easier to start, since the starter-generator only has to spin up the core engine; the power turbine and propeller can come along in their own sweet time. Moreover, they tend to be quieter (again, especially on the ground), since the prop and power turbine can run considerably slower than the gas generator group. On the other hand, they're considerably harder to start in the air on their own power, that is, battery starts work fine, but "windmilling" starts are dicey on some, while they tend to work better for direct-coupled engines.

Temperature and Turboprops

Operationally, both types are very similar, and both types have very similar control systems, although some free-turbine installations offer the pilot an extra control, it's usually a set-it-and-forget it type.

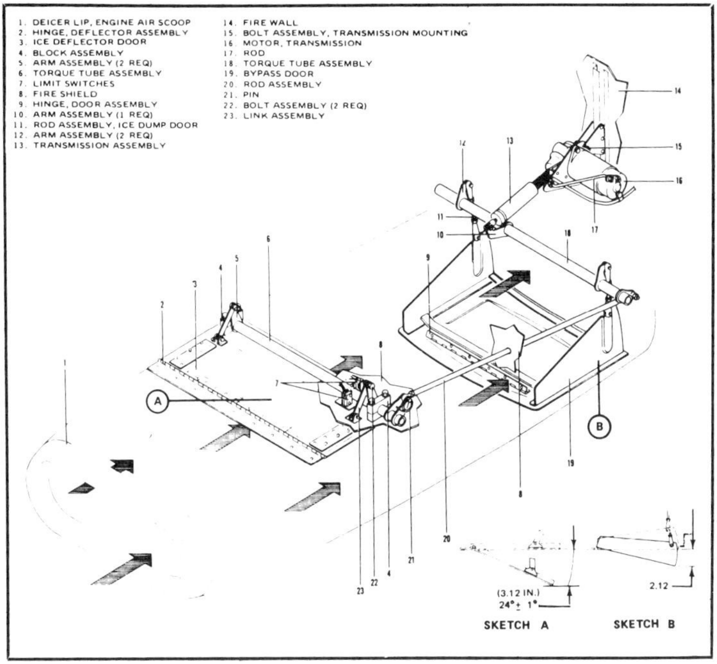

Fig. 10-7. Ice protection for PT-6 engines is provided by an "inertial separator" system. With the ice deflector door extended and the aft bypass door open, incoming air must make a sharp turn to enter the engine, while ice particles continue in a straight line and are dumped overboard through the bypass door. In this installation (Piper Cheyenne), a small electric motor operates both doors; in King Airs, a manual handle does the same job.

What about a turboprop's operating limitations? Well, they're expressed a bit differently from those of a recip, but many of them mean the same thing.

Temperature is what's really critical for turboprops. Overtemping a turboprop is worse by far than overboosting a recip (although you can "overboost" a turboprop as well). Although actual combustion temperatures are similar for both types of engine, a turboprop is far less tolerant of excessive heat.

Why is this? Largely because of a combination of two factors: individual component size, and the continuous-combustion nature of turboprop operations.

Let's look at the hot parts of a recip for a moment. Sure, it gets plenty warm in those cylinders, but consider one fact for a moment: a cylinder is a pretty massive chunk of metal, and while it's hot inside, it's cool outside; plenty of heat can be dumped through all those little cooling fins with which it's adorned. Moreover, it operates in close proximity to other big hunks of metal, and can transfer a lot of heat to them.

Even more important, that cylinder operates *cyclically*. It gets hot in there during the combustion stroke, and remains fairly warm during the exhaust stroke, but then there's a chance to rest and cool off while a fresh charge of nice, cool, unburned fuel and air are brought in during the intake stroke and compressed during the compression stroke. Thus, the cylinder spends less that half its time getting heated up, and the rest of its time cooling down.

Now let's look at that poor little turbine blade down in the hot section of that turboprop engine. First of all, it's *little*, perhaps less than an inch long, and only a quarter inch wide and perhaps a sixteenth of an inch thick. It's out there all by itself, too. It has no big cylinder with lots of cooling fins attached to it. True, it can pass a little heat on to its base where it's joined onto the turbine wheel proper, and where the oil in the bearings will carry some away, but other than that, it's surrounded by nothing but air, which is an insulator.

It's worse than that, though, because it isn't surrounded by air, it's surrounded by hot combustion gases, the same ones that heated it up in the first place, and *they're there all the time*! That single fact is what makes a turbine engine so temperature-sensitive. It never gets a chance to cool down, so its metal is exposed to much greater thermal stress than that of a recip engine burning its fuel at the same combustion temperatures.

Oddly enough, the highest temperatures don't always come when the engine is working hardest. Often, they come during engine starts, when combustion has already been established, but before there's sufficient airflow through the engine to provide any cooling at all. Moreover, it puts More stress on engine parts to be heated up than to remain hot once heated, just like recips. Most engine manufacturers recognize this by allowing *momentary* excursions above normal redline during start, but they also show the detrimental effects of heating and cooling in the engine by limiting some engines overhaul periods, not only by hours, but by a number of start-shutdown *cycles*. Even with the current cost of fuel, engine overhauls cost more, that's one reason you'll often see commuter airlines continue to run the engine on the side away from the door while emplaning or deplaining passengers. They'll keep the number of cycles on one engine to a minimum; then, after a while, they'll switch it to the other side of the airplane.

If temperature is a major limiting factor in turbine operations, there's another parameter that's a primary one for setting power and that is the *torque*, or the force with which the engine is turning the output shaft.

In fact, torque *is* power; it's a direct measurement of how hard the engine is working. Why don't we use it, then, instead of manifold pressure as a power parameter for recips? As a matter of fact, in big airline radials, we do. But, for the smaller recips, it's simpler, more practical, and cheaper to measure manifold pressure, and it's almost as accurate.

The temperature at which a turbine engine operates depends to a considerable extent (more so than for a recip) on the temperature of the intake air. This leads to a situation in which an engine's limiting parameters may change with outside conditions. At higher temperatures, an engine wil run hotter to produce a given torque output. By the same token, at lower temperatures, torque for a given fuel flow increases. The same holds true for increased air *density*; thus, for most turboprops, torque is a limiting factor at low altitudes and/or temperatures, and temperature becomes limiting at higher ones. At some point, there's a crossover. Operationally, you can see it. Take off on a cold day, and you'll have to stop the power levers well back of the full *throttle* position as torque nears the redline. As you climb, you'll have to advance them further and further, just like a nonturbocharged recip, and as you do, the temperature indicator will be climbing. At some point, you'll be near temperature redline, and from here on you can't advance the power levers anymore without overtemping. This is the crossover point; as you climb higher from here, torque will start to drop off.

Power Controls and Instrumentation

This leads, naturally, enough, into a discussion of turboprop power controls and instrumentation. Your primary control is a *power lever*. It's analogous to a throttle, but it does quite a bit more than just change fuel flow. It is connected to the engine fuel control unit, and does control fuel flow to control power, but in many installations, it's also connected to the propeller pitch governor, so that adding power not only increases fuel flow, but propeller pitch as well (just like the throttle on a recip when the propeller is in a normal governing range).

If we're dealing with a free-turbine engine, we'll also find a propeller control, functioning exactly like the one on a recip. However, one vital function of prop control not found on most recips is still controlled by the power lever, and that is the *reverse thrust*.

Actually, there's even a range between forward power and actual reverse, called the *beta* range. In normal forward-thrust operation, propeller speed is controlled by the prop governor, which adjusts blade pitch to absorb whatever power is generated by the engine. In beta range, movement of the power lever *aft* of the normal idle position *directly controls propeller pitch,* while an *underspeed governor* keeps the engine running at the proper speed. Beta operation is normally restricted to the ground. In beta, the pilot can continuously vary the net thrust from forward, through none at all, to slightly reversed. Further aft movement of the power lever (after the propellers have already gone into reverse pitch) increases fuel flow once again, thus increasing reverse thrust.

There's one more control for most turboprop engines, whether free of direct coupled turbines, and that is the *condition lever*. This does different things for different engines.

For free turbines (such as the PT-6, one of the main functions of the condition lever is to set idle speed. The aftmost position of the lever is a complete fuel cutoff, and is used for stopping the engine. Further forward are two detented positions—low idle and high idle.

These positions do nothing more than their name implies—they set the speed at which the engine idles. Low idle (generally around 50-60 percent of redline speed) is used for ground operations, while high idle (75-85 percent) is preferred for use as a flight idle position.

Actually, it's not all that critical for most operations. With the power levers forward of their flight idle position, moving the condition levers to either position has no effect. However, in some installations, windmilling drag is significantly greater in low idle, and it may take longer to shift from forward to reverse thrust on landing if the engines are allowed to "spool down." On the other hand, low idle is quieter and smoother for taxiing. On a free-turbine engine, the windmilling drag won't be much different from that of a recip (if anything, it'll be less) regardless of condition lever position or idle speed. Old King Airs, which didn't have reverse thrust, had no condition lever at all. They ran in low idle mode all the time, and worked fine!

Operation of the condition lever becomes somewhat more critical in the case of direct-coupled engines, and now wind-milling performance *does* become a factor. Geared directly to the compressor of a gas turbine, with its huge power demands, the windmilling prop of a turboprop has *much more drag* than that of a recip. So much more, in fact, that most aircraft using direct-coupled engines incorporate some sort of autofeathering (or negative-torque-sensing) system. In such airplanes, retarding the throttles below a *flight idle* setting could result in massive drag and associated high sink rates, so one function of the condition lever is to keep the engines spooled up while in the air.

Another is in lieu of a propeller control; there's no separate prop handle on a direct-coupled engine. Instead, the condition lever typically offers a low idle, or *start and taxi* position set for somewhere around 60-70% rpm, a full-rpm *takeoff, climb, and land* setting, and a *cruise* detent, typically at about 96% rpm. This sounds unusually high, but think for a moment—a recip with a 2500-rpm redline would have a 96% rpm cruise of 2400 rpm, which is an entirely plausible figure. As with the free-turbine engine, the condition lever also has a fuel-cutoff position. In this case, it is used primarily for emergency shutdowns, and linked with a feathering valve in the prop governor.

What about this instrumentation? Let's start with a temperature indicator. In direct-coupled engines, this gauge reads exhaust-gas temperature, just like in a recip. In free turbine types (and a few direct-coupled ones as well), it gets its signal from a pickup between two stages of the turbine, usually between the compressor turbine and the power turbine, and reads *Interstage Turbine Temperature,* or ITT. In either case, it provides a readout of combustion-gas temperature at a critical point, and is the prime instrument to monitor to avoid overtempting during a start or inflight.

Torque is measured by an oil-pressure arrangement incorporated in the reduction-gear mechanism. It can serve as either a power-setting parameter or an operational limit, depending on conditions. It's analogous to the manifold pressure gauge on a recip.

Next, we'll need a *tachometer*. These are usually calibrated in percent of maximum allowable rpm, rather than in actual turns per minute, probably because the marks would have to be very close together if you wanted to get the range of 0 to 50,000 rpm on one scale! Often, this is a dual-scale instrument with a large needle that sweeps over the marks from 0 to 100%, while a smaller inner dial goes once around every 10 percent for fine readout.

If we're running a free turbine, we'll need *two* tachometers—one as described above for gas generator rpm, often also called N_1 or N_g rpm, and another, more conventional type for setting propeller rpm with the prop control.

Finally, let's install a *fuel flow indicator*. Since turbine power output is directly related to how much fuel it's burning, this is a very useful power setting aid. Remember that there's no real "throttle" in a turbine, since its air passages are fully open all the time. The only way we can control its speed and power is by contrtolling how much fuel it burns, add more fuel, and it'll run faster. Mixture control is more or less automatic, dictated largely by the turbine's air-moving capabilities. Give it too much fuel, and it'll start running too hot; give it too little, and it'll slow down until it gets to the point at which the flame in the combustion chambers gets weaker and finally blows out altogether.

A word of explanation here: why does the temperature *increase* when we have too rich a fuel mixture (too much fuel flow for a given airflow), rather than decreasing as it would in a recip?

This is because a turbine "runs lean" compared to a recip—typically, only 20 percent of the airflow supports combustion, while the remaining 80 percent is for cooling. Thus, we're operating on the "lean side of peak;" enriching the mixture will cause EGT (or ITT) to increase.

This is also the phenomenon responsible for those high temperatures on starting—the fuel flow "gets ahead" of the slowly increasing airflow. Obviously, the bigger and more powerful a starter we have, the cooler a start we can anticipate. This is also why we leave the starter engaged until the turbine is well on its way toward idle rpm, perhaps as high as 50% rpm or more. We want to help it through the hot low-rpm area as quickly as possible.

Turboprop Propeller Systems

With their ability to deliver reverse thrust, as well as their need to operate at fairly high speeds even on the ground, turboprop powerplants,

Fig. 10-8. Garrett TPE Turbine engine (courtesy of Beechcraft).

particularly the direct-coupled kind, need rather sophisticated propeller and governor systems. We've already touched briefly on the beta range; now, let's take a more detailed look at the components and operating modes for a typical direct-coupled engine (we'll discuss the free-turbine type in our subsequent pilot report of a King Air).

In fact, we may as well look at one of the most sophisticated installations currently available, that of the Cessna Conquest. In addition to the normal under and overspeed governors and the prop governor, the Conquest's Garrett TPE-331-8-401S engines incorporate an electronic fuel control computer that can relieve much of the pilot's engine-monitoring workload (Figs. 10-8, 10-9 and 10-10). With the fuel computer in operation, the power levers can be advanced all the way for takeoff without danger of overtempting or overtorquing the engines.

The engine can be operated in two modes: normal, in which the fuel computer is operational, and manual, a degraded mode that does not use the electronic fuel computer. In manual mode, the condition lever is inoperative (except for its emergency feather and shutoff function), and the engine runs at 100% rpm at all times. Fuel flow, torque, and temperature are controlled by the pilot with the power level. Reverse thrust is also unavailable in manual mode.

In addition to the fuel control, various other governors are included in the system (Fig. 10-11).

The *propeller governor* is a more or less conventional unit which controls propeller, and hence engine, speed during normal forward thrust operation. When the power levers are at or above flight idle, the propeller governor is set by the condition lever for the desired speed, which can range from 96% to 100%. With the power levers below flight idle (ground idle, beta range, or reverse), the propeller governor is inoperative.

The *underspeed governor* controls engine speed directly, not by prop pitch, like the prop governor, when the power levers are anywhere other than the range between flight idle and minimum power. Its minimum setting is 65% rpm, and its maximum is about 96%, above which the prop governor takes over. The underspeed governor is also controlled by the condition lever, but only when the power lever is below the flight idle position.

On certain installations, there's also an *overspeed governor* which limits rpm in case of failure of the propeller governor.

The electronic fuel control incorporates an automatic engine start sequencing system which senses engine speed and performs various

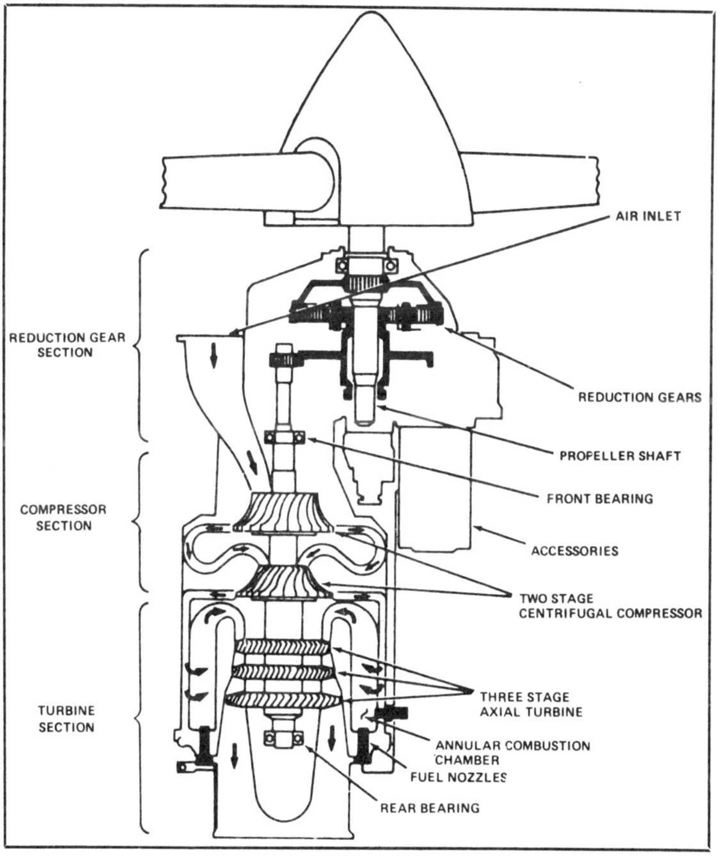

Fig. 10-9. A schematic of the Garrett TPE-331 engine shows that there is no separate gas generator and power turbine; the turbine section drives both compressor and propeller reduction gearing with a single shaft.

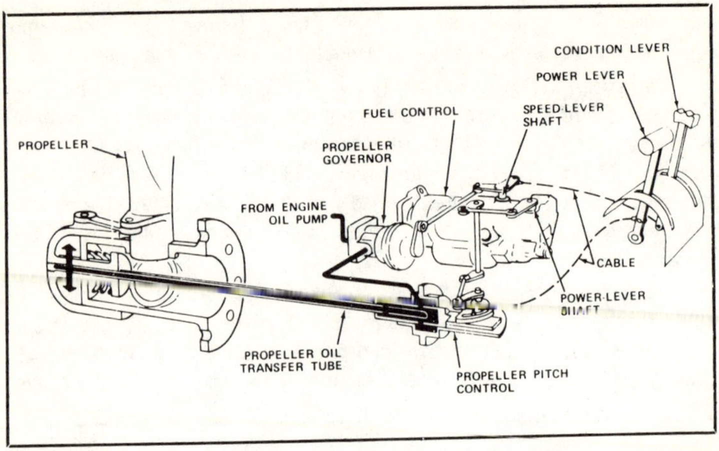

Fig. 10-10. In a Garrett engine installation, the condition, or RPM, lever controls engine RPM via the fuel control (at idle or in "beta" range) and via the propeller governor (governing range). The power lever controls engine fuel flow via the fuel control, and directly controls propeller pitch in "beta" range via the propeller pitch control (not the prop governor).

switching functions automatically. These include starter energizing ignition and fuel control, and generator reset and start. When operating in manual mode, these functions must be switched by the pilot.

Finally, there's a *Negative Torque Sensing* system (NTS). This system reduces pilot workload during an engine failure situation by sensing negative torque valves (in other words, situations when the propeller is driving the engine rather than vice versa), and alleviating them by dumping oil pressure from the propeller, allowing it to start moving toward the feathered position. NTS isn't quite the same as an autofeather system, since the blades won't go all the way toward feather, but will halt at some point at which the negative torque value drops below the NYS threshold, but it has nearly the same effect. It greatly reduces the drag of a windmilling prop, which can then be completely feathered (and the engine shut down at the same time) by the pilot's pulling the condition lever fully aft to the *emergency feather and engine shutdown* position. (Normal engine shutdown is controlled electrically by the fuel computer unit.)

Let's follow one of these engines through a typical start, taxi, takeoff, climb, cruise, descent, landing, and shutdown procedure.

Starting the Turboprop

You'll recall that a smooth, rapid start is the key to avoiding dangerous overtemperatures. That starter-generator unit really has its work cut out for it. It must accelerate the engine *and propeller* to about 10% rpm before the fuel and ignition are cut in, then continue helping the start and acceleration process until the 60% mark. Even a "mere" 10% is over 4,000

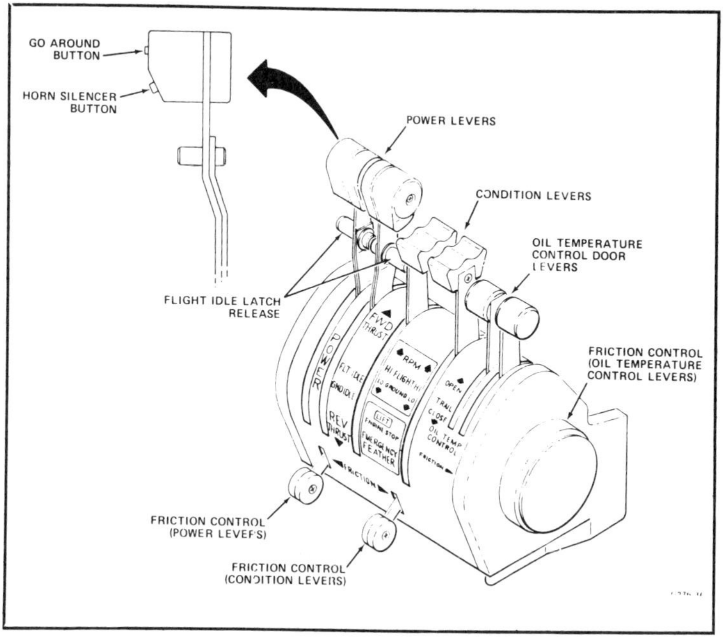

Fig. 10-11. Here's the power quadrant of a Rockwell Commander; note that the "condition" levers (RPM levers would be a better name) also have engine cutoff and prop feathering functions. In other aircraft, these are handled by a separate control. The power levers control engine fuel flow in forward-thrust settings, and also control prop pitch in the range just below "FLT IDLE" to full "REV THRUST."

rpm, while 60% is 24,000 rpm, which is a heady speed indeed! Obviously, this requires a lot of electrical power; turbine aircraft put a tremendous strain on their batteries, which must usually recharge for a minute or so after the first start before the second engine can be started. If a ground power unit is available, it's a good idea to use it even on warm days, and it's just about mandatory on cold ones, remember, the more juice we can shove into that starter-generator, the cooler the start! (Some airplanes have two batteries and the capability of running them in parallel for normal start, or in series on a really cold day, to throw 48 volts across the starters. A series start with fresh batteries looks quite impressive, although it's hard on the starter-generators; you get the impression that fully-charged batteries would give you enough "oomph" to fly once around the pattern on starter power alone!)

You'll also recall that anything we can do to make it easier for the starter-generator to accelerate the engine and get as much air as possible moving through it helps the situation. One thing that's mandatory is to make sure the prop is in as flat a pitch as possible, and this is where the "start locks" come in.

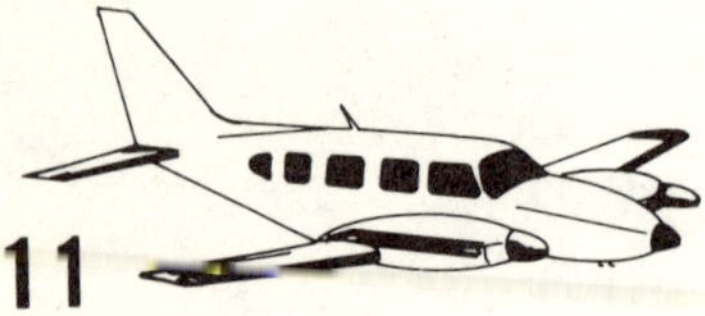

11

Flying A Typical Turboprop: The Beechcraft E90

There are a number of reasons for including a pilot report on the venerable Beech King Air as an introduction to turboprops. For one thing, the last chapter dealt almost entirely with direct-coupled engines, so a look at the Beech, with its free turbine PT-6 engines, will familiarize you with the differences in operating that type of powerplant. Moreover, if, when, and as you move up into a "Kerosene Queen," chances are good that it'll be a King Air. This plane was the first small turboprop on the scene, and it's been a popular one ever since, so there are an awful lot of them out there!

The history of the King Air dates back to the early 1960's (actually, to the late 1950's, when the Army expressed interest in a small turboprop for utility, VIP transport, air ambulance, and electronic-surveillance missions). Beech already had an airplane about the right size, but powered by recips, known as the Queen Air, which was a roomy medium twin based on the wings (and, originally, the tail) of the old model 50 Twin Bonanza, and powered by a pair of the dread Lycoming IGSO-380 mechanical supercharged maintenance hogs, each putting out some 340 (some models 380) hp, when they worked.

Providentially, at about the same time, United Aircraft of Canada (soon to be part of Pratt and Whitney) came out with one of the first of the small turboprop engines, the PT-6 which was about the same physical size as the Lyc, but it put out some 500 hp. It was a marriage made, if not in heaven, at least in Wichita, and the King Air was born.

Ironically enough, the airplane the Army ended up getting wasn't quite the King Air. Their mission, as it turned out, didn't call for pressurization, and besides, this was in the long-lost days when the military paid at least token attention to what things cost. As a result, the Army version, still flying today as the UC-21, is basically just a Queen Air with PT-6 powerplants. Beech, however, continued with the development of the airplane, and soon the first civilian King Airs appeared, with a similar, but pressurized, fuselage and a pair of 500-hp PT-6's purring away on the wings.

Fig. 11-1. Beechcraft Super King Air 200 is a 289-kt eight-place craft (courtesy of Beechcraft).

The early birds were a little crude compared to their descendants. For example, their pressurization system was run hydraulically from the left engine, rather than by bleed air, and reversing props weren't standard (although most buyers opted for them at extra cost), but they were still among the flagships of the corporate fleet, and a lot of them are still around. I spent some time flying the 14th King Air built only a couple of years ago, and it was an excellent airplane. In fact, as this is written (1979), a good early King Air with reverse, spiffy interior, and good avionics, including radar and flight director, would probably sell for under $250,000, quite a bit less than a new, but smaller and less capable, light pressurized twin with recip powerplants.

Enough sales talk, though; it is sufficient to say that the original King Air 90 was the progenitor of a current model line that includes the C90, or "plain Vanilla" King Air; the E90 (same airframe, bigger engines); the Super F90 (slightly bigger engines yet, same fuselage, T-tail); the A100 (longer fuselage, even bigger engines—still PT-6's, though, like all the other so far-dual-wheel landing gear, conventional tail); the B100 (an A100 with direct-coupled Garrett engines in case you're weird or something and don't like PT-6's); and the Super King Air 200 (Fig. 11-1, 11-2 and 11-3),

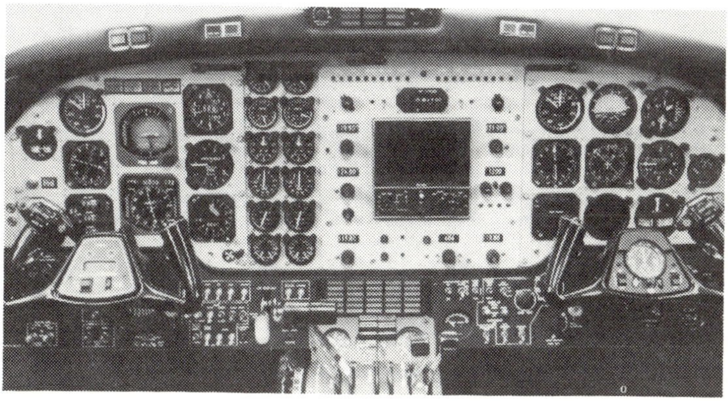

Fig. 11-2. Super King Air instrument panel (courtesy of Beechcraft).

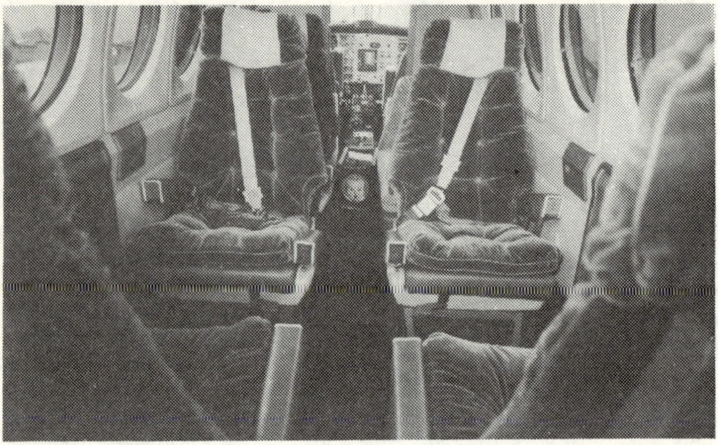

Fig. 11-3. Super King Air cabin (courtesy of Beechcraft).

basically an uprated A100 with T-tail. Confused? So am I. For this pilot report, let's pick a middle-of-the road airplane, say, an E90 (Fig. 11-4), of the most popular models.

The Turboprop Preflight

Preflight offers no hidden surprises; there's the usual collection of fuel drains here and there on the bottom of the wing and under the wing and under the wingpits near the fuselage. There are six fuel fillers—the two mains at the wingtips, for 196 gallons in each wing; two aux fillers on the wing center section, with 41 gallons per side; and a pair of fillers for the nacelle tanks, high on the back of each nacelle. Normally, these are used for maintenance only, since the nacelles fill by gravity from the mains. In fact, if you open a nacelle filler with the mains full, you'll be greeted by a veritable gusher of jet fuel, so don't even try them.

Jet fuel can harbor considerably more water than a gas, so drain the sumps carefully. If flight above the freezing level is planned, you should be sure that fuel onboard contains the necessary concentration of anti-icing additive. Many fuel suppliers blend this in as a matter of course nowadays; if your supplier doesn't, the additive can be blended during fueling, use "hi-flow Prist." If you're planning operation in out-of the way areas, it's a good idea to carry a few cans along.

The engines have the distinctive PT-6 appearance—props feathered when at rest, and big exhaust pipes right behind the spinners. That's because the PT-6 is a reverse-flow engine, although the air intake scoop is at the front of the nacelle, the actual engine intake is almost all the way at the back, and the direction of engine airflow is forward, with the exhausts just behind the propeller gearcase. As the air travels down the duct behind the scoop, it passes the inertial ice separator vanes, which should be retracted at this point. Just below the air intake scoop is another, smaller, scoop for the oil cooler. If this is the first flight of the day, the scoops will most likely be covered. Now's the time to pull all covers and give the prop a gentle twirl, listening for any unpleasant noises from the gearcase.

The whole side of the nacelle swings up for engine inspection, giving an excellent view of all components, linkages, and so forth. The oil dipstick is at the rear of the engine on the accessory section at about 11 o'clock. If you're new to turboprops, you'll be surprised at how clean the oil is, it's hard to see on the dipstick.

High on either side of the nose is a louvered opening. This is for the electrically-driven Freon air conditioning system. The airplane has electric heat, too, and while some may decry such systems as old-fashioned, they have one great advantage: you can summon a ground power unit (the plug is under the right wing on the outboard side of the nacelle), connect it, and preheat or precool the cabin. The environmental system needs about 300 amps at 28 volts. Starting loads are about the same after an initial pulse of around 1000 amps.

Checking Out the Interior

With a thorough external preflight completed, we can open the airstair door, if it's not already open, of course, aided by a damper that lets it down gently. The lock mechanism includes two pins on each side and a pair of hooks at the top, and the pilot's manual cautions that if the handle isn't moved all the way over the center to the locked position, pressurization loads on the hooks will rotate the whole mechanism and open the door. To prevent such an unfortunate turn of events, there's a positive lock button that drops out when the handle is locked, and inspection windows in the door allow you to see all four pins. Moreover, when the airplane is pressurized, a bellows mechanism interposes an additional lock plunger, preventing the handle from moving.

Various cabin arrangements are available, ranging from club seating to couches along the sides. If a potty is installed, it's also a legal belted seat.

Fig. 11-4. Beechcraft King Air E90 is one of six King Air models. High cruising speed is 249 kts, and service ceiling is 30,910 ft. (courtesy of Beechcraft).

Fig. 11-5. Beechcraft King Air E90 instrument panel (courtesy of Beechcraft).

The cockpit is well laid out for single pilot operation, and a good many King Airs are flown that way. In addition to the two flight groups, engine instruments run in two vertical rows down the left side of the center panel, while the rest of the center area is given over to radios and the radar indicator (Figs. 11-5 and 11-6). Electrical switches are on the subpanel, arranged by groups from left to right—first the battery and generator switches (together with the loadmeters), next the engine starter and ignition switches. Just below the panel between the pilot's knees are the two pull-handles for ice-separation vanes.

Ligically enough, the other ice-protection switches are next along the bottom of the panel, including pitot heat, prop de-ice, hot windshield engine fuel control heat (normally used at all times, even in non-icing conditions), stall warning, fuel vents, and electrically heated boots around the edges of the engine intake. Above them are switches for prop governor testing, which we'll describe in a moment, and the autofeather system; above these are lighting switches.

Directly below the radio stack are the three green lights for the gear, the flap position indicator, and the cabin altitude/differential pressure and cabin rate of climb indicator. Further right are the gear handle, a bunch of switches and knobs for the environmental control system, and most of the circuit breakers.

A good-sized throttle console runs aft between the seats, starting with the power, prop, and condition levers and a big elevator trim wheel on its left side. Further aft along it are the flap control, rudder and aileron trim wheels, and the pressurization controller. Then there's a bit of empty space, often used for additional radio equipment, and finally a couple of coffee cup holders which always fill me with trepidation, because just behind them, on the aft face of the pedestal, are the most critical circuit breakers. I shudder to think of what might happen if a cup of coffee were to be spilled and find its way into the back of that breaker panel.

A comprehensive annunicator panel runs along the glareshield, which often includes an angle-of-attack indicator right in front of the pilot. The only controls in the overhead are a bank of cockpit lighting controls, which can be preset, then turned on and off all at once by a master lighting switch. A small side panel on the copilot's side contains pneumatic system gauges and the oxygen gauge, as well as some fuses for certain instruments.

The pilot's side panel contains the fuel system controls, and that's where we turn our attention as we prepare to start the engines and get underway.

The Turboprop Fuel System

The King Air fuel system is a simple one (Fig. 11-7). Each side of the airplane has a main tank and an aux tank. The aux feeds the main via an ejector-type transfer pump run by fuel flow from the engine. There are only two fuel gauges, one for each side, but they can be switched to read either main or aux fuel (Fig. 11-8). In addition a float switch in each aux tank illuminates an *aux tank empty* light on the fuel panel, alerting the pilot to turn off the "aux transfer" switch for that side.

Each main tank contains an electric standby boost pump which can be used in case the engine-driven boost pump fails. The standby pumps are also required for cross-feeding operation, which is achieved by opening the cross-feed valve and turning on the boost pump on the side *from* which you want to feed.

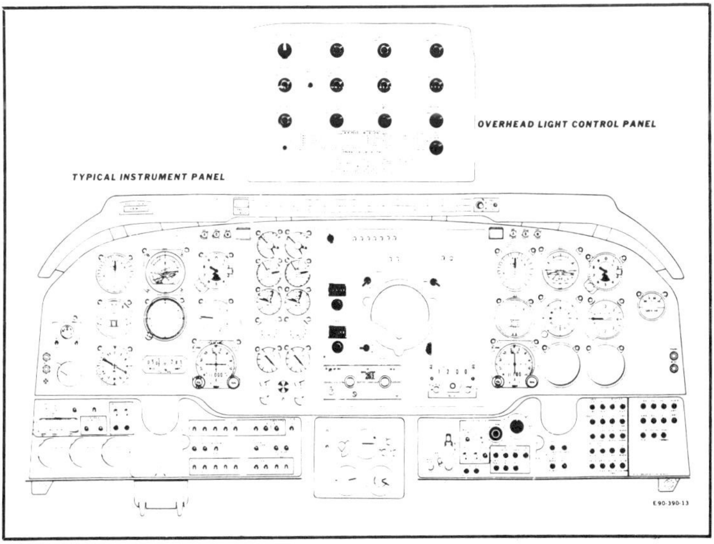

Fig. 11-6. The standard arrangement of the King Air panel puts the radar and radios in the center; two rows of engine instruments flank the pilot's flight instrument cluster. Flap position and pressurization indicators are on the center panel.

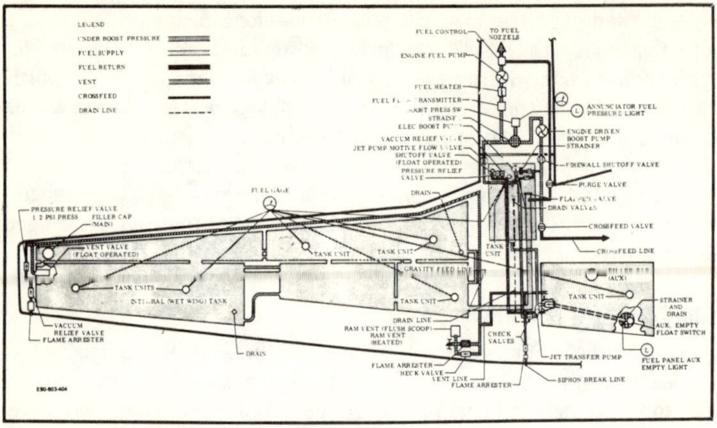

Fig. 11-7. Although this King Air fuel system incorporates a number of tanks, they all act as a single large tank, feeding via gravity and fuel-powered jet pumps into the nacelle tank and thence to the engine.

You might note that all the fuel system functions (including the firewall shutoff valves) are electrically actuated or controlled. You may be wondering what would happen in the event of a total electrical failure. A provision has been made for this—each vital component has a dual electrical supply, both from the aircraft's normal busses and from a *hot battery bus* which is live at all times. Diodes prevent reverse current flow (in fact, they're used extensively throughout the electrical system; in most conditions, a generator failure will result in the remaining generator assuming all electrical loads with no pilot workload required, as there are no manual bus tie breakers to be closed) between the normal and "hot battery" systems. It's a simple, reliable, and effective arrangement. Perhaps its only disadvantage is that the pilot must ensure that all pumps are turned off before leaving the aircraft after shutdown, since they'll run even with the master off, and rapidly deplete the battery.

Engine-Start Sequence

Let's begin the engine-start sequence, then, with the fuel panel, even before turning on the master. That way, we can verify that it's getting power from the hot battery bus. After going through the usual cockpit check to ensure that all the knobs, handles, and switches are in the right position, we'll turn the two firewall shutoff switches (after lifting their guards) to "closed." Next, we'll turn the standby boost pumps "on" and listen to make sure they're running.

Now, we'll turn on the battery switch, and take a look at the annunciator panel. Among the various lights, we should see that both engines show low fuel pressure. Since the standby pumps are still on, we know that the firewall shutoff valves had, indeed, closed properly. If we open them now, the fuel pressure lights should go out. Now we'll shut off one standby boost pump (the low pressure light for that side will come on again) and open the crossfeed, which should make the light go out once again. Next, we'll turn the idle boost pump back on, and the other one off (low fuel pressure lights should both stay out), then we'll close the crossfeed and verify that the light on the side of the inoperative pump comes back on. This sounds involved, but the whole process only takes about 10 seconds, and we've now verified several important things: first, we know that both boost pumps work; also, we know that the crossfeed valve works properly; plus, we are sure that the crossfeed itself works in both directions.

Finally, before going on the actual start, we'll push the "volts" button next to each loadmeter in turn (Fig. 11-9). They should both indicate the same, verifying that the current limiters (an expensive way of saying fuses,) that tie the left and right main buses to the battery, are intact. A zero indication would indicate that the current limiter on that side was open, which is the equivalent to an open (and non-resettable) bus tie circuit breaker.

Let's fire up: Normally, the right engine is started first, since it's closer to the battery. We'll check that the prop control is full forward, that the condition lever is at cutoff, and that the power lever is at idle. Fuel boost pumps are not required; in fact, in normal operation, the engine-driven boost pumps can handle all loads.

After clearing the prop area, we'll flip the right starter and ignition switch on, and the engine will start to wind up. The prop will barely move

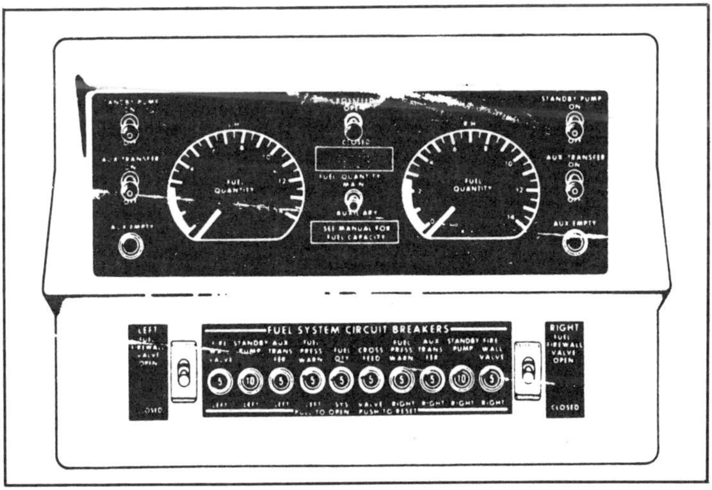

Fig. 11-8. The simple fuel panel of a King Air includes gauges that can be switched to indicate quantity in the (inboard) aux fuel tanks. Fuel is transferred to the mains by aux transfer pumps, selected next to the gauges; a light indicates when transfer is complete. Vital electrical functions for fuel control are energized even with the aircraft master turned off; be sure fuel pumps are off before leaving the aircraft, or you'll return to a dead battery!

at first, but remember, these are free-turbine engines, and the gas generator section will accelerate rapidly, as monitored by the N_1 tachometer in the engine stack. When it stabilizes, usually around 15% but not less than 12% unless we really want a hot start, we'll move the condition lever to *lo idle*, keeping our hand on it ready for an immediate cutoff if the temperature looks like it's heading for the start limit.

That seldom happens, though (at least, as long as the battery is in good shape), and the engine lights off at once (indicated by a rise in ITT, Interstage Turbine Temperature) and continues to accelerate. As it reaches 50% or so, we'll turn off the starter/ignition switch. Oil pressure should be monitored during start, but there's a bit of a problem in doing that. The oil pressure gauge is run by 26 volt AC, and will remain dead until we turn on an inverter. However, we have an excellent oil-pressure reference in the prop itself—if it comes out of feather, there must be oil pressure!

Ready to start the next engine? First, let's set the condition lever of the right one to *hi idle* and turn on the right generator. If you watch closely as you turn on a genearator, you'll see a slight rise in ITT, that's the fuel control unit adding a bit of power to keep the engine on the correct idle speed as the load of the generator is applied. You'll also see a warning light on the annunicator panel, indicating a high rate of battery charge. After all, we've just worked the battery pretty hard starting that the first engine. Normally, this light should go out within a few minutes after engine start. If it stays on more than 5 minutes, it's a sign that all is not well with the NiCd battery, and we should shut down and check it out.

With the generator helping the battery, the left engine will start a bit faster (and a few degrees cooler) than the right one. The sequence is the same. Switch on, monitor N_1 rpm, condition lever to "lo idle," monitor temperature. As the left engine comes up to idle, we'll cut off the starter, turn its generator on, and bring the right engine back to "lo idle" as well. Finally, we'll turn on both inverters (to make sure they're both with us), then leave the selector on one of them, and the fuel flow, torque, oil pressure, and oil temperature gauges will wake up (as well as all the 115 VAC gadgets such as flight director, RMI, radar altimeter, and so forth).

At this point, too, we can turn on all the avionics and the fuel control heaters. We'll also turn on the cabin environmental control system. This last item can take quite a bit of electric power, so this is a good time for a look at the loadmeters, N_1 rpm, and ITT again. Normal operations of the cabin system is in either *auto heat* or *auto cool* modes, with a thermostat switching the appropriate components on and off as required; manual override is provided if necessary. In the auto mode, the system provides two controllers, one in the cabin and one on the flight deck. A switch selects which one will be used, depending on personal preference and employer-employee relationships. (" . . . deny me a cost-of-living increase, will he? Hah . . . I'll roast him to a turn . . .")

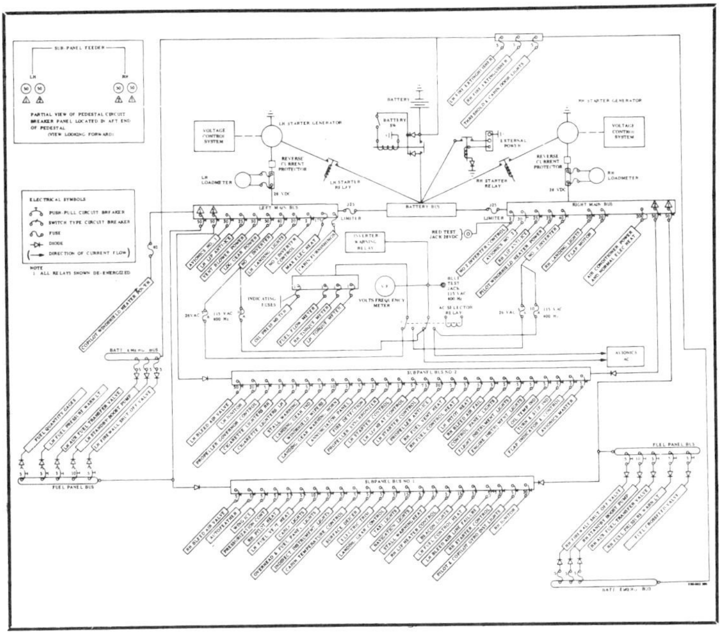

Fig. 11-9. The electrical system of the E90 King Air is separated into several busses, interconnected by current limiters (fuses) and diodes (small arrow symbols). The diodes allow current to flow only in the direction of the arrow, providing automatic fault isolation. Certain functions (fuel shutoff valves, pumps, fire extinguishers) are fed both by the normal system and by an always-powered battery emergency bus, so they can be operated even with the rest of the electrical system shut down.

Preparing for Takeoff

Taxiing out, we'll check brakes and instruments, then halt in the runup area. A real *runup* isn't always required, but there are a few items that should be checked at least on the first flight of the day, and we'll detail them here.

Most of them have to do with the reversing propeller installation and its associated governors and safeguards. First, we'll check the overspeed governor. Normally, this is set to restrictmax prop speed to 2200 rpm; however, a test switch on the panel momentarily resets in to 1900 rpm, so we'll slide the condition levers up to "hi idle," hold the test switch, and carefully work the power levers forward. Prop speed should stop increasing at 1900 rpm, while further forward movement of the power levers (it only takes a bit) should result in torque and ITT increasing. Then, we'll return the power levers to idle.

The next test is at the other end of the prop range, and is to check operation of the *secondary low pitch stop*. Not all King Airs have this device, the newest ones don't, and it's been removed from some of the older ones.

The purpose of the secondary low pitch stop is to guard against inadvertent prop reversing in flight. If the prop were to go past the normal low pitch limit, it would soon come up against the electrically-activated secondary stop, halting further pitch decrease and lighting a warning on the annunciator panel.

In normal operation, the secondary stop is disengaged electrically by connections to the power lever when they're pulled below idle by the pilot. For test purposes, a switch disables this function, so that the props actually do come against the stops and light the indicator. To carry out the test, we'll start by noting the prop rpm at high idle, then we'll depress the test switch and pull the power levers back toward reverse until the low pitch stop lights come on (the handbook says "align the aft edge of the power lever with the top of the reverse marking," but rigging can change; most pilots I know just start easing back until the lights come on.) Prop rpm should stabilize at around 200 above high idle, and *must increase* when the test switches are released.

Next, let's have a look at the autofeather system. Normally, it will only arm completely at fairly high power settings, but yet another of the ubiquitous test switches can be used to check it out at low power.

With the engines idling, we'll move the autofeather switch to the test position. For the moment, the two *autofeather armed* lights on the annunciator should remain out. However, as we advance the power levers to around half power (500 ft/lbs torque), the system should arm, and the two green lights will come on.

Now, let's slowly retard one of the power levers. As we bring it back, the first thing that will happen is that the *other engine's* autofeather armed light will go out. The engines "watch each other," and either one will only feather if it thinks its buddy on the other side can carry on alone. (Actually, this is to prevent both engines feathering if the system were left armed and power was reduced for a descent or landing.) As we retard the power lever further, the *armed* light on the engine we're retarding will go out and it will start to feather. Now, we'll bring the power back up and check the other side the same way.

The system actually monitors engine torque, and in the test mode it'll never feather completely. As the prop nears feather, the torque required to keep it turning will increase, and it'll unfeather again, then just oscillate back and forth. In normal mode, of course, there would be little or no torque available, and it would feather all the way.

Finally, let's check manual feather by simply pulling both prop controls all the way back and feathering the engines. This sounds horrifying if you're used to recips, but the PT-6 doesn't mind; ITT and torque will increase somewhat, but the engine will just sit there, spinning away with a feathered prop. In fact, it's a courteous thing to do if there's some light aircraft waiting behind you. However, keep an eye on oil temperature, since the oil cooler depends on a certain amount of prop blast for best effectiveness. (Do it only at low idle, by the way.)

Turboprop Takeoff

That about takes care of the props; now, we'll check ignition and ice protection, and then we can fly.

The *engine autoignition* switches are on the pilot's subpanel; like the autofeather system, the ignition "watches" torque. We'll advance the power levers until we see around 400 ft/lbs, then turn on the autoignition; green "armed" lights indicate that the system is ready. If engine torque goes below 400 ft/lbs, the ignition for the affected engine will come on, lighting a yellow "ignition on" light and extinguishing the "armed" indicator. This system can be tested at the same time as the autofeather, if so desired.

Finally, we'll pull on the ice protection vanes; a slight drop in torque indicates proper operation. Unless we're going to launch into icing conditions, we'll retract them again to get maximum power for takeoff.

Handling of the condition levers on takeoff is, to some extent, a matter of taste. Once the power levers go forward into a flght range, the engines will spool up to an appropriate rpm regardless of condition of lever position. From a passenger viewpoint, things do seem a bit smoother if power application occurs in one continuous buildup, rather than two distinct steps (first an rpm increase as the condition levers go to "hi idle," then another as the power levers are advanced.

I generally start the takeoff roll with the condition levers at low idle, but advance them to high idle as soon as the power levers are brought up. All flight operations are normally carried out with condition levers at the high idle setting.

As soon as the airplane leaves the gound, a switch on the landing gear starts the pressurization system working. To avoid a sudden inrush of bleed air that could get ahead of the outflow valves, causing a temporary "ear bump," the system energizes the two bleed valves separately, with a six-second delay between them.

As we transition to climb power, there are a few additional items to remember. This is the time to turn on the prop sync system and turn off the autofeathering switches, since we're now high enough above terrain that we'll want the additional help of a windmilling prop in trying to airstart a failed engine. After a few moments, this is also a good time to flip the fuel gauges over to *aux tanks* and make sure they're feeding properly.

As in any other airplane of this class, there's little actual aircraft management required in cruise; leaving the crew ample time to perform navigational tasks and flight management duties.

Descents, however, do have to be planned slightly in advance. The PT-6 powerplants will continue to put out adequate bleed air for pressurization at high idle power, but if you've forgotten and left the condition levers at low idle, your ears may let you know as you begin your letdown.

Turboprop Emergency Procedures

Before we get into the land phase, let's discuss a couple of typical King Air emergency procedures, not that these represent common occurrences, but they'll give you an idea of the sort of things you might have to deal with someday in real life.

If an engine fails in cruise on a recip, it'll keep windmilling. All you have to do is provide the proper combination of fuel, air, and ignition and it will relight.

To some extent, the same is true of direct-coupled turbine engine. Since the prop keeps windmilling and everything is coupled to the same shaft, the engine will continue to turn (at a higher rpm than the starter could probably achieve), and restarting is a matter of providing fuel, air, and ignition, just like in a recip. The only difference is that the turbine's compressor has a lot more drag than the pistons of a recip, so the windmilling prop will have a lot more adverse effect on the aircraft than that of a recip would have. This is why many direct-coupled engines have NTS or autofeather.

In the case of a free turbine, though, it's a bit different. Depending on how it failed and the airspeed and altitude of the aircraft, the engine may windmill, or it may not.

If the compressor and turbine of the PT-6 actually come to a stop, of course, the prop will feather whether the autofeather system is armed or not, and regardless of prop control position. This is because the prop governor depends on engine oil pressure for operation. In such a case, the only possible airstart is a starter-assisted one, which is essentially no different from a ground start (although the engine and prop will accelerate somewhat faster once they get going.)

In many cases, however, the combination of ram air blowing into the engine's compressor and *suction* caused by the still-spinning *power* turbine (the only turbine actually linked by gearing to the propeller will keep the N_1 end of the engine spinning fast enough to maintain adequate oil pressure, preventing the prop from feathering, and a *pure* or unassisted airstart is possible. Once again, it's a matter of bringing the power lever back to idle and making sure that the igniters are on, then monitoring ITT as the engine comes back to life and accelerates. Normally, with a well-charged battery, an assisted start will still be "the cool way to go."

Let's look at one more adnormal situation, *a landing gear malfunction*. Since the King Air remains more or less unique among the turboprops in using a simple, all-electric, mechanically-actuated landing gear, a direct descendant in concept, if not in detail design, of that of its Beech ancestor, the Bonanza.

Actually, the backup system uses some components of the primary unit (linkages and bell-cranks), but has its own source of power—your right arm. Pulling up a handle on the floor unclutches the gear system from its normal electric motor and connects it, instead, to a pump-like red handle next to the seat. Actually, this handle is merely a ratchet drive like a gigantic socket wrench and pumping it up and down moves sprockets and chains (shades of the Wright Brothers and their bicycle shop!) to extend and lock the gear. The system isn't designed to retract the wheels;

attempts to do so are likely to damage it, and leave you with the gear struck half up and half down.

On the other hand, there is something you can do with the system that's difficult or impossible with most others—you can practice an emergency extension. Just pull the gear circuit breaker, put the normal gear lever down (nothing will happen), and you can then pull the emergency clutch lever and pump away for all you're worth. Once the gear is down, just push the clutch back to its normal position, stow the "pump" handle, push the circuit breaker back in and you can retract and extend the gear normally.

Landing a Turboprop

Landing a King Air is another area where there are various schools of thought in regard to power management for maximum smoothness combined with effective reversing performance.

We can consider it a foregone conclusion that the condition levers will be at high idle during the approach and landing. Remember, once again, that once the engines are developing appreciable *forward* thrust, the condition levers have no effect. If you get bored while cruising along, you can move them from high to low idle, or even have one high and one low, with no apparent effect.

On landing, though, you're going to be bringing the levers back through idle into beta and/or reverse, so once you're on the runway, what you do with the condition levers might be influenced by how much runway you have and how much reverse thrust you plan to use (if any).

For any use of reverse,the props must start out in low pitch, that is, with the prop handles forward. Check before moving them, however that the "secondary low pitch stop" light is not illuminated. If it, it indicates that the normal low pitch mechanism is defective, and you shouldn't attempt to use reverse on that engine. (Incidentally, if a prop ever does reverse in flight, your first indication should range from a change in engine note to the airplane unexpectedly swapping ends, depending on how much forward power you were carrying and whether the engine that reversed inadvertently maintained that power. Those who have had, and survived, inflight reversals have generally gotten out of the situation by rapidly identifying and immediately feathering the bad engine.)

Very well, the props must be full forward if we plan to reverse after landing. The problem is that if we move them forward on final, the resultant "mmmmrrRROOOAWRRrrr" is likely to unnerve the folks in the back. A technique with a bit more finesse is to leave the props at cruise rpm until after touchdown. A *reverse not ready* light on the annunciator panel will remind you of this fact, and on some airplanes, a solenoid wired to the same circuitry keeps the power levers from being moved aft of idle until the light is out, then, on touchdown, bring the power levers to idle, the props full forward, and get on with the business of reversing. With the power levers at idle, prop lever movement won't change RPM appreciably.

What about the condition levers? You'll have more useable reverse thrust, and it'll be useable sooner, if you leave them in high idle. Otherwise, engine rpm will start to drop as you come back with the powers levers, then spool up again as you get into reverse. Many pilots, including myself, leave the conditions in high idle until the power levers are definitely in reverse (noticeable torque and ITT rise), then slide them to low idle before coming out of reverse, at which time the engines will smoothly decelerate to taxi rpm.

After taxiing in, there's not much to shutting down. It is a good idea to be sure that the ITT's are below 550 deg C or so to reduce thermal stress on the engines. With NiCd batteries, you can also check charging rate to be sure you're going to have enough electricity for a nice cool start next time around. This is best checked by turning off all electrical equipment and noting the remaining loadmeter reading. A low load reading indicates a well-charged battery.

For most rapid engine spooldown, handy when you're going to be deplaning passengers, you can pop the props into feather before pulling the condition levers to the cutoff position. If you don't the props will feather anyway, but they won't have a *head start* and it'll take a bit longer.

Finally, before leaving the airplane, make sure no boost pumps are running. Remember, they don't need the battery switch on to run. Forget them, and you won't have a hot start next time, in fact, you won't have any start at all, since the battery will be completely dead! It'll most likely have to come out of the airplane, too; even if you get a GPU start, the battery will have *deep cycled*. It'll have a tremendous charge rate as soon as you get an engine started and a generator on the line, and that's a great way to get a thermal runaway.

Dealing with a Dead Battery

In keeping with this, I'll close the chapter with a final word on what to do if you ever *are* stuck out in the backwoods with a dead battery, but can start the airplane with a GPU. Here's how to get things going with the least risk of themal runaway, although the risk is there. This practice is hard on generators, and will cost you some time and fuel, but someday you may be stuck, and it'll get you home. (The Voice of Bitter Experience says: "You felt dumb when you left the master of a 150 on overnight when you were a student? Kid, you don't know what dumb feels like until you've done it with a King Air...")

What we want to do is start an engine, and then start charging the battery as slowly as possible. As soon as a generator comes on the line, it'll put out as much juice as it's capable of providing. What we have to do is try to trickle the electricity into the battery gently, rather than dumping it in all at once.

First of all, then, let's plug in the external power cart. (This assumes that it's one of the simple ones common at most airports; some of the really sophisticated ones can provide only a trickle of electricity themselves if necessary, and can charge a battery quite gently on their own.)

Next, get all your ducks in a row for the start so you can crank with a minimum of delay after you turn on the aircraft battery switch. You want to reduce the length of time that the GPU is pouring all its power into your battery before you get the starter on the line to take the load itself.

Then, start the engine. As soon as it's self-sustaining, have the GPU removed.

Now, get ready to turn on the generator. Try to do it at a nice low N_1 speed, for two reasons: first you don't want to suddenly jam a bunch of current into the battery, and second, as soon as the generator is turned on, it'll "see" the heavy load of a discharged battery; this, of course, makes the generator harder to turn. If you turn it on at too high an N_1 speed, you can actually break its shaft where it's geared to the engine accessory section. (Many starter-generators have an intentional "weak link" in the shaft, so that if something *does* get damaged, it's the starter-generator rather than the engine.

Finally, as soon as the generator comes online, reduce the amount of current going to the battery. You can do this by draining it off elsewhere, or by running other electrical loads. Monitor the loadmeter so you don't exceed the generator's capability, and turn on avionics, blowers, landing lights; whatever will do the job—electric heat or air conditioning is wonderful. After you've established a load figure, watch carefully for a few minutes; if it continues to decrease, the battery is charging properly. If it *increases*, the battery is running away. Take it off the line at once, shut down the airplane, and check the battery compartment with a fire extinguisher at hand.

The battery will have to be removed from the aircraft and replaced or overhauled and bench-charged in a qualified NiCd battery shop, before further attempts are made to start.

12

Stability Augmentation, Flight Control And Aircraft Monitoring Systems

As the sophistication and capabilities of the aircraft we fly have increased, their avionics have kept place. Some of the most conspicuous advances have come, especially in recent years, in the areas of navigation and flight management, and we'll cover those in greater detail in the next chapter.

Now, however, we'll look at what might be called "airframe" avionics; those black boxes that don't necessarily have that much to do with the navigational situation, but rather with the basics of flying the airplane itself. Such avionics can include stability augmentation systems (devices that make the airplane handle better and/or more predictably and safely); flight control systems, usually grouped under the generic name of "autopilots;" and systems and performance monitoring systems, a catch-all category that might include units as diverse as radar altimeters, angle-of-attack indicators, and "smart" fuel flow indicators.

Understanding SAS

Let's start with SAS. To those unfamiliar with the subject, those three letters conjure up images of transpolar flights, with icy Aquavit served by even icier nordic stewardesses. Unfortunately, in our work-a-day world those letters stand for Stability Augmentation System, and SAS is already standard on few turboprops and is likely to work its way downward through lighter and lighter aircraft in coming years.

In a nutshell, any SAS is a device or combination of devices designed to improve the stability, handling characteristics (also called "flying qualities"), and safety of a given aircraft. In a sense, it may appear that older aircraft are better than new ones, since they seem to get by quite adequately without any sort of SAS, but this isn't quite the case.

You see, as advances continue to be made in the field of basic aerodynamics, various new and innovative aircraft configurations are discovered and developed which can offer hitherto unreachable levels of performance and efficiency. Such advances, however, are sometimes only realized at the expense of flying qualities. Consider something as simple as an aircraft's tail, for example; it contributes virtually nothing to the ship's performance *per se*, but must nonetheless be of a certain size (with, of course, a certain weight and a certain drag) to provide adequate stability. If the tail were smaller, the pilot would have to constantly manipulate the controls, making tiny corrections to overcome the aircraft's natural tendency to wander away from straight and level.

How about, though, making the tail smaller and hooking up some sort of black box to make the constant little corrections? That, in essence, is the philosophy behind SAS in its various forms and functions. It probably reaches its highest development at this writing in such aircraft as the F-16 and F-18 jet fighters, which utilize completely, computerized electronic "fly-by-wire" control systems. There's no manual reversion, either; even if manual controls were installed, the basic aircraft is, for all purposes, unflyable by a mere human—it's just too unstable and unpredictable. Anyway, that's what ejection seats are for . . . (In all fairness, it should be noted that the computer systems are installed with quadruple redundancy. After all, the Air Force doesn't have to pay for them, you and I do . . .)

Note that when I say "fly-by-wire" I don't mean that the pilot is just along for the ride—he or she still controls the airplane with a more or less conventional stick and rudder pedals. However, there's no direct connection between the stick and the control actuators; instead, it all goes through the computer. The result, from a piloting standpoint, is that the SAS system isn't "there" at all. The airplane simply responds naturally and predictably to all control inputs. In fact, in just about any aircraft, you won't notice a SAS until it fails.

Obviously, systems such as those in the F-16 are still pretty heady stuff for airplanes such as those we'll be flying. Nonetheless, some SAS units are already in use, the most common being those on the Piper Cheyenne II and III aircraft. These are rather simple units whose primary function is to change the pitch "feel" of the airplane by positioning a bungee in the elevator system in response to angle-of-attack inputs, and the airplane's tail is, of course, large enough for it to be flown without SAS. Even so, there's enough of a control-response change that no-SAS flight is only permitted within certain airspeed, weight, power, or CG restrictions.

Moreover, there's one very basic form of SAS that's been around for years on a large number of twins: the *yaw damper*. This is nothing more than a system to keep the aircraft coordinated, using a rudder servo and either a yaw sensor (rate gyro) or side-force sensor (accelerometer). Yaw dampers are often offered as options to many of the popular autopilot systems, and they come highly recommended. Even when they are not absolutely necessary for safety, they can go far indeed toward improving overall ride qualities and passenger comfort. This becomes particularly important in a couple of specific situations. For example, some aircraft are especially prone to *Dutch roll*, a type of roll/yaw coupling that interchanges energy between roll and yaw modes. The result is an uneasy, cork-screwing wing-rock tendency, particularly in rough air, that not only makes instrument flying a chore, but can soon have even hardened passengers doing the "Technicolor yawn."

Another area in which yaw dampers are handy is in larger (i.e. longer) airplanes, such as the Swearingen Merlin IV/Metro series (sometimes described as "the tunnel at the end of the tunnel.) In an airplane with passenger seats far aft of the CG, the last few riders get swished around pretty hard in turbulence unless there's a yaw damper, and if there's a long nose, the same applies to the pilot.

Staying on Top of Your Autopilot

The next step beyond SAS is actual flight path control, as performed by the ever-popular autopilot. Bascially, there's nothing an autopilot does that a human instrument pilot doesn't do; it's just that the autopilot has quicker reflexes, doesn't get tired, and has a much higher threshold of boredom than even the most hardened reader of this book. Autopilots have come a long way in the corporate and commuter segments of general aviation. After starting out as hissing, clanking hydraulic monsters that weighed a hundred pounds or more and could do little more than hold the airplane more or less straight and level (when they worked at all), they've evolved through only slightly smaller, but more accurate and reliable, early electric types replete with banks of evilly-glowing vacuum tubes, to today's lightweight all-solid-state units that are not only able to fly the airplane to any desired attitude, but can hold altitude and heading, track ground navaids, and even shoot (and, if necessary, miss) approaches down to the lowest CAT II ILS minimums. Moreover, these systems have considerably more control authority than the older ones, as well as offering approach precision far surpassing that attainable by the best human pilots.

With all this, modern autopilot systems are both easier and more complex to work that older ones. Easier, in that they incorporate a number of features to reduce pilot workload. For instance, most of them offer automatic pitch sync, which means that they can be engaged in any attitude, and they'll "take over" with no change in the flight pattern (with older systems, the airplane often had to be precisely leveled and trimmed before the system could be engaged). Automatic trim is another virtually universal feature. The autopilot "feels" how much control force it has to apply, and trims this force out of the pitch axis just as you would, so that whenever you disengage it you're given back an aircraft that's properly in trim. (In some older airlines systems, cockpit drill was for both crewmembers to hold on the yokes when the autopilot was to be disengaged, in case it was massively out of trim!)

Moreover, the various and sundry automatic features are becoming easier and easier to use. For example, with the push of a button, the system can be commanded to fly a given heading unitl intercepting a given VOR course, then turn inbound or outbound as required and track that course, making appropriate crosswind corrections. This doesn't sound difficult on the surface, and it's something that we human pilots do every day, but remember—flying a heading, intercepting, and tracking each take a different set of *control laws*. We humans switch back and forth between them unconsciously; the autopilot has to "know" each set, "know" when it's appropriate to switch from one "technique" to another, and make the transition smoothly and appropriately.

At the same time, operation of modern autopilot systems is a bit more demanding than old ones. Not so much the actual system operation, of course; it's just that, as the autopilot becomes more and more sophisticated, it gets harder and harder for the human pilot to stay ahead of it.

Autopilot programming changes must be made in the right way, and at the right time, and the system must be monitored constantly for proper operation. The more functions and capabilities an autopilot has, the more buttons there are to push, and the greater the chance of hitting the wrong one!

There's an insidious danger, too. With a really good autopilot onboard, many pilots fall into the habit of engaging the unit and then putting their brains on "hold" until they have to hand-fly again. This can be a rough situation indeed if the autopilot blows a fuse or otherwise has to be disengaged ahead of schedule, leaving the pilot miles behind the airplane and the situation.

There's another reason for staying ahead of the autopilot, too. Even with the increased authority of modern units, they do have their limits, and it's not all that hard to present them with a situation that they can't handle very well.

An ideal example for this is the ILS at my home town, Santa Barbara, California. The usual *feeder fix* is the Gaviota VORTAC. From there, the route extends down the 163 degree radial some 6 nm to a right-angle intersection(!) with the localizer ("HABUT" intersection).

Now, a right-angle localizer intercept is no joke under the best of conditions. If you're going at any sort of speed, the only quick way to intercept the localizer is to execute the time-honored *bank and yank* technique the instant the localizer needle unpegs from full deflection.

Of course, an autopilot cannot be nearly as brusque as a human pilot. Most autopilots are limited to bank angles of 30 degress or less, and roll into turns fairly gently.

Compounding the problem at Santa Barbara is the fact that Gaviota VORTAC sits on top of a respectable range of hills just inland from the coast. Although the VORTAC can be crossed as low as 3500 ft MSL going the other way, ATC in its infinite wisdom usually runs incoming ILS customers over the station at 6000 feet MSL. The IAF altitude at HABUT, only about 6 nm away, is 3500 feet, meaning that you'll have to lose 2500 feet in only 6 nm, or about 420 feet per mile. At 150 knots, that means a descent rate of almost 1100 fpm.

An Autopilot-Controlled Approach

Let's assume, now, that we're going to shoot a coupled (autopilot-controlled) ILS approach to SBA. We've crossed Gaviota at 6000 feet, and are on our way down to 3500 feet, heading toward HABUT. Here's where it's important to stay ahead of the autopilot and the situation. If we cross the localizer at 150 knots (remember, we're hitting it at exactly 90 degrees!), we'll sail more than two nm past it by the time the autopilot, with its limited authority, has us turned around and headed back toward it. It'll cut back in at 45 degrees, cross the localizer again, and finally settle down with a series of diminishing S-turns; we'll be lucky if the autopilot has gotten its ducks in a row by the time we hit the outer marker.

No, what we have to do is help the autopilot as much as possible by slowing the airplane early and giving it plenty of lead on descents and turns.

This means that we should slow to some speed, such as final approach speed, *before we even get to Gaviota*. As we cross the station, back with the throttles (and, in a recip, out go the wheels and flaps to let us keep enough power to keep the engines warm), and down we go. We want to get to 3500 feet as expeditiously as is safe, while avoiding airspeed increases that will make it hard to intercept the localizer.

As soon as we have the descent underway, we can turn our attention to the autopilot horizontal modes. (I'm assuming that we have a modern unit and takes its course datum and deviation information from the pilot's Horizontal Situation Indicator, but the technique is the same for older types that use a heading "bug" on the DG.)

Tracking inbound to GVO (Gaviota), we were most likely using the VOR mode of the autopilot, and we'll normally set it for the 163 degree outbound radial as we crossed the station. As soon as we see that the autopilot is tracking properly outbound, let's note the *heading* it's holding (which includes any possible wind corrections), then set that heading on the heading bug and switch the autopilot to heading mode. True, we're now flying a simple mag heading, rather than actually tracking the outbound radial, but we're flying the heading that the autopilot was using for its correct track, and we're not going to get that far off course in 6 nm. Switching to heading mode frees the course arrow, so we can set it to the inbound ILS course, at the same time tuning and identifying the SBA ILS on the #1 NAV receiver.

As we near 3500 feet, we'll be ready to level off by engaging altitude hold and, possibly, retracting the landing gear. We'll add power if necessary, but remember, we want to stay nice and slow, because that localizer's going to go by pretty fast. This is an area where modern units with auto sync and auto trim are really appreciated. All it takes to level off from a climb or descent is the push of a single button, with an occasional glance at the airspeed to make sure that the power setting is appropriate.

If we have the sort of autopilot that allows intercepts on any (pilot-chosen) heading, this is the time to engage (or, rather "arm") the approach mode. The airplane will continue on the 163-degree terminal route heading until the localizer needle unpegs, then turn to the inbound course to intercept it. With older autopilots, you'll have to watch that localizer needle like a hawk; as soon as it quivers, turn the heading bug (which, on these older units, is also a course datum) to the inboard course and select "approach" mode.

In either case, what we're doing is giving the autopilot a chance to get intimate the with localizer as early as possible; and, given our relatively low airspeed, it has a good chance of nailing the inbound localizer with a minimum of overshoot and subsequent S-turning.

Now, as it happens, 3500 feet at HABUT is just about on the glide-slope, and if all things were as they should be in this best of all possible worlds, you could just lock onto the glideslope there and "slide down the wire" all the way past the outer marker (6.5 nm down the road) and on to the airport. Unfortunately, this isn't allowed under current rules, since there's a mandatory descent to 1900 feet after crossing HABUT inbound. (As far as I can tell, this is primarily for traffic separation, and may well be eliminated once SBA's new radar goes into service in mid-1980.)

Whatever the reason, the mandatory descent is not altogether a bad thing, since many autopilots will only capture the glideslope from below, in fact, one of the most popular has to "see" a pegged "fly-up" needle for 20 seconds to arm the glideslope function.

Again, it's back with the power, and down we go, to get a fly-up indication and get down to 1900 feet as fast as we can. Whether or not to extend the gear at this point can vary from airplane to airplane. In most of the ships I fly in and out of SBA, I find that by flying level with gear up and approach flaps down as far as the outer marker, and then extending the gear at the blue light, I can achieve the right rate of descent on the glideslope with only minimal power changes.

Once we're down at 1900 feet well outside the outer marker, and slowed to a reasonable approach speed, the worst is over. As we get closer to the marker, we'll see the glideslope needle start to come down. As it centers, the glideslope mode of the autopilot becomes active, and we'll start down the final approach segment.

If this sort of flying sounds too mechanized for you, if you'd rather do it yourself, there's still a system that makes it possible for you to enjoy all the benefits of the autopilot's computational capabilities while still flying the airplane by hand.

The Flight Director System

This is the Flight Director system, and its basics are almost identical to those of an autopilot. In fact, the only difference is in what is done with the system output. In the autopilot, system outputs run servo units that position the controls, while with a flight director, the outputs run tiny miniaturized servos that position *command bars* in the pilot's artificial horizon, now called an ADI (Attitude Director Indicator).

In fact, with exception of command bars vs. servos, the rest of the system is identical, so much so that most systems installed nowadays are so-called *Integrated Flight Control Systems*, which offer both capabilities.

At any time, the pilot can elect to follow the command bars by hand, or can engage the autopilot with the flip of a switch to do the same thing.

Presentations within the ADI differ among manufacturers, but there are two basic styles. One is a pair of crossed needles like those in a ILS indicator. However, these do not show raw ILS data, but rather computed steering and pitch commands. For example, if the system commands a 30 degree turn to the left, the roll steering needle will deflect to the left and remain there until the airplane has been banked to the required amount, then it will center and remain centered while the airplane turns. As it nears the desired heading, the needle will deflect to the right to command the rollout, then center once again as level flight is reached on the new heading.

The other form of presentation is the so-called *V-bar*, or *single-cut type*. Instead of the two cue needles of the first type, V-bar units present the pilot with a shallow inverted "V" symbol, while the horizon's traditional "miniature airplane" has been replaced by a delta-winged symbol. To follow flight director commands, the pilot need only keep the miniature airplane "tucked in" the V-bar as it pitches up and down and rolls left and right.

Using a flight director, especially during low approaches, can be a real boon. The flight director vastly simplifies instrument scanning, since all the information you need is presented on the ADI, or is it?

Well, it is if everything is going well. Remember, though, that a centered V-bar or centered needles don't necessarily mean that you're on course; they just mean that you're satisfying the system's commands. You could be well off course (correcting back), and the needles would still be centered, so even when using and enjoying a flight director, make sure you keen an eye on the raw data of course, altitude, ILS position, and airspeed to monitor the system.

Remember, too, that most flight director computers will not command greater angles or rates than those appropriate to an autopilot. Thus, with a flight director, you have the option of monitoring raw data and, if necessary, exceeding the flight director commands for faster intercepts or closer tracking. The system will never lead you in the wrong direction, but it may sometimes not lead you in the right one as fast as you may want.

Monitoring and Controlling Devices

A final area of *airframe avionics* is that which deals with devices to monitor and control aircraft performance and the flight path. These can include power-control systems such as the electronic fuel-control system on the Cessna Conquest's Garrett engines (discussed in an earlier chapter), fuel-system monitors, angle-of-attack indicators, and radar altimeters.

Let's discuss the last-named instrument first, since it falls into this category more or less by default (I couldn't figure out where else to put it).

A radar altimeter is nothing more than a little radar set mounted on the underside of the airplane which bounces its pulses from the ground below. Rather than a radarscope, the readout is on a panel indicator, typically calibrated from 0 to 2500 feet or so. The scale is usually expanded to read very accurately up to, say, 500 feet, then slightly compressed up to 1000 feet and considerably compressed beyond that. Thus, the instrument can be read very accurately in the all-important lower ranges.

The primary use for the radar altimeter is as an approach aid, since it provides highly accurate height information during the final phases of an approach. This information is much more accurate (and, of course, much more easily interpreted) than that provided by the pneumatic altimeter. Using the pneumatic altimeter, you have to fix a different figure in your mind for the DH or MDA for each new approach. With a radar altimeter, you just set the warning "bug" for the appropriate height *above ground*, and lights will flash and bells ring when you reach that height. No mental calculation is required—400 feet is 400 feet whether you're at Denver or at Death Valley.

Moreover, the pneumatic altimeter generally lags behind the actual situation, especially with large rates of descent. Not so the radar altimeter; it's always up to date.

A radar altimeter can be more useful than just an approach aid, however. Whenever I have one, I've gotten into the habit of setting the warning bug at 2500 feet on climbout. This, of course, makes it sound off on descent, reminding me to reset it for the DH or MDA. However, it also provides me with a sort of "poor man's ground proximity warning system" at no extra charge. For instance, let's say I'm cruising along and I haven't noticed that my altitude is gradually slipping, when I get to within 2500 feet of terrain, bells and whistles. It's a good safeguard in mountainous terrain, too. It takes a pretty steep mountain to go from 2500 feet below you to above your altitude in less space than you take up turning away from it.

If the radar altimeter is a good approach aid, an angle-of-attack indicator is a terrific one for landings and takeoffs. Optimum landing performance, after all, isn't necessarily obtained at one specific airspeed, (you've seen how that can change with flap setting and/or CG and/or aircraft weight). What *is* important is a specific angle of attack, and if you had some sort of instrument to read that, you could virtually ignore airspeed. Remember, too, that like the altimeter, the airspeed indicator lags behind the true state of affairs, while angle-of-attack indicators are more or less instantaneous.

While angle-of-attack indicators are more common on larger aircraft (Fig. 12-1), they're available in the aftermarket for anything down to the light singles, and they greatly simplify landing airspeed management. One of the most popular has only three marks on its face— a *low diamond* for short field landings at about 1.3 V_{so}, a *barber pole* for normal landings, and a *high diamond* for landings in gusty conditions (which is also near best rate-of-climb speed for departures). As long as you fly the airplane to put the pointer on the appropriate mark, you *know* that you're at the optimum angle of attack, and hence the optimum airspeed, no matter what that may be. It's very instructive to shoot a few landings with one of these gadgets under varying conditions. For a given angle of attack, airspeed can change considerably with aircraft weight and/or configuration. It'll also give you a vivid indication if your stall speed goes up in even a gentle bank. Since the indicator shows, in effect, margin above the stall, you'll be surprised at how much you have to speed up in turns to keep the pointer on the "barber pole."

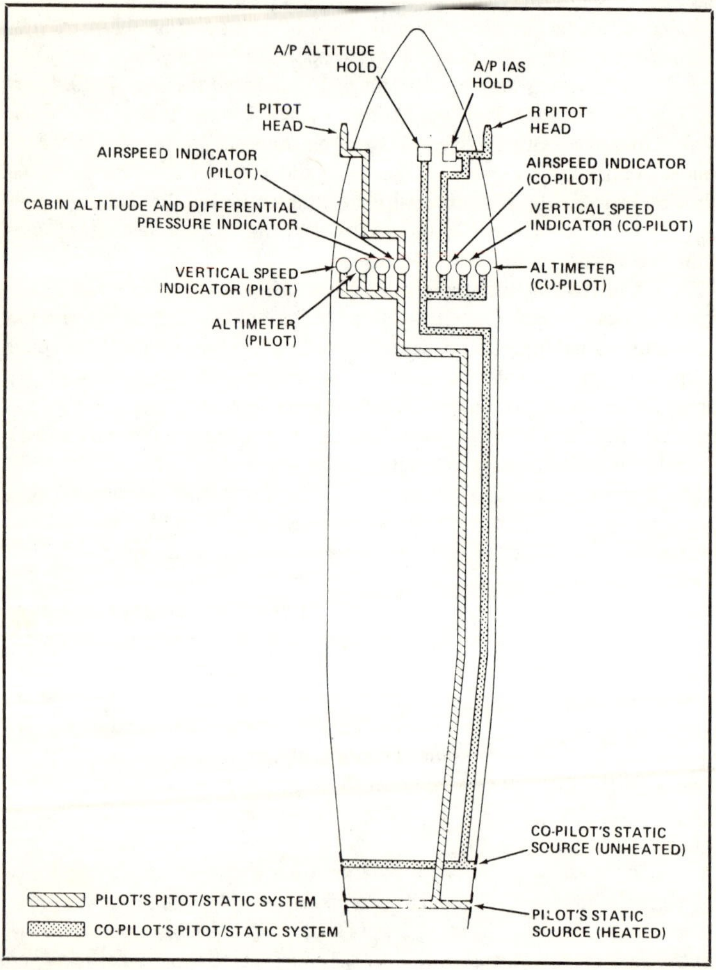

Fig. 12-1. Most larger twins have separate pitot and static systems to operate the pilot's and copilot's instrumentation. In this Rockwell Commander system, the pilot's system also provides a reference for pressurization instruments, while the copilot system is used for the autopilot's altitude hold function.

While the simpler angle-of-attack indicator systems are only effective in the low-speed regime, some turboprops have full-range indicators which are also accurate in climb and at cruise. Using one of these you can fly the airplane to specific points for best rate of climb, best endurance (for holding), best range (for cruise economy), and so forth. As fuel becomes more and more expensive, such systems can "pay for themselves" in fewer and fewer hours.

Since we're on the subject of fuel, let's consider a final new item of airframe avionics: electronic fuel-flow and fuel-quantity indication.

To understand the advantages of such systems, it'll help to look at their predecessors, the systems that are installed in most new airplanes today. Fuel quantity is measured in the tanks, either with simple floats or with more accurate capacitance probes, but all systems have errors. Fuel flow is measured with pressure or flow transducers in the powerplant fuel lines. Reciprocating-engine types are particularly inaccurate, since they're basically just pressure gauges, easily misled by clogged fuel injectors (see the chapter on recip engines). Turbine types, using analog flow sensors are somewhat more accurate.

With the advent of microprocessors, or "computors on a chip," the situation has changed. Now, it's possible to use simple and reliable *digital* flow sensors that literally measure the fuel passing through them "drop by drop," with much greater accuracy. Connect such a sensor with a suitably programmed microprocessor, one incorporating an accurate time signal (quartz clock), and you can "count the drops" for a number of purposes.

For example, compare "drop count" with time, and you get a direct, and very accurate, readout of fuel flow. Compare the total number of drops you've counted with the amount of fuel known to be onboard at the beginning of the flight, and you have a very accurate indication of both fuel consumed and fuel remaining.

Moreover, the computational capability of the processor allows other added features. For example, it can display fuel consumed or remaining in either pounds or gallons at the flick of a switch. Flick another, and it compares fuel remaining with current flow rate to provide a readout of *flying time remaining* at the current power setting. This can be awfully reassuring to know in an extended holding situation, as can the knowledge of how much you have aboard down to the nearest pound of gas!

13 Long Range Nav And Flight Management Systems (RNAV)

The most spectacular advances in avionics in the last few years have come in the areas of navigation and flight management. Ten years ago, the progression of technical advance was usually from the military to the airlines to general aviation. By now, that's been reversed to the extent that there are a few medium twins flying around that can do things that even the 747's can't!

The basic airway structure of the United States is pretty well defined by over 800 VOR's , most of them incorporating DME as well. While it's a pretty good system, the population of aircraft using it has far exceeded the expectations of its designers, with the result that the avoinics industry is continually searching for better ways to deal with its limitations, or to operate outside them altogether.

Currently, the most common advanced navigation system used in general aviation aircraft is the so-called RNAV, or *area naviagation*, system. The RNAV abbreviation actually stands for randome navigation, and is something of a misnomer. Since area nav depends on signals from existing VORTAC's and since most units operate by "displacing" such stations electronically and moving them to the desired track, few RNAV systems are actually "random." (Actual "random navigation" has also been defined as "the track flown by a student pilot on his/her first solo cross country between the points of departure and landing, not necessarily at the planned destination.")

RNAV's Functions

What an RNAV set actually does is relatively simple in this day and age when every toaster-oven hs its own microprocessor—it does elementary grade-school level trigonometry. If you'll cast your minds back, you may recall that, given two sides and the included angle of a triangle, the remaining side and the other two angles can be determined. That's all an RNAV does.

Normal RNAV use involves planning a cross-country flight, defining the track in terms of *waypoints*. A waypoint is, in effect, a VORTAC that has been "moved" electronically away from its actual location to one more convenient for the flight at hand. The parameters defining a waypoint, then, are its distance and direction from the "parent" VORTAC.

That distance is one side of the triangle we mentioned above. The RNAV set "knows" this distance, since it's been set in by the pilot, and it also has the bearing from the station to the waypoint.

With the station tuned in once airborne, the ship's VOR and DME supply the RNAV set with the aircraft's distance and bearing from the *station.* Now, the RNAV set has everything it needs to calculate the remaining side of the triangle, which is the distance from the aircraft to the *waypoint,* while the bearing is calculated in similar fashion. It's nothing you douldn't do with a pocket calculator or slide rule (or, for that matter, with chart, plotter, and pencil), but the RNAV set does it much more conveniently. It also drives the aircraft navigational displays and the autopilot. The FAA has published numerous route structures and instrument approaches based on the use of RNAV equipment.

The first RNAV sets were simple affairs that could work with only one waypoint at a time. While some current units can still only display a single waypoint at a time, most can store the parameters for a whole series of waypoints, so that they can be programmed before takeoff when pilot workload is at a minimum.

The most sophisticated units not only store the basic waypoint parameters of distance and bearing (and, in some cases, VORTAC station elevation for slant range correction); they can also store the frequency of each defining VORTAC, and will automatically tune the NAV and DME receivers to the correct station when each waypoint is selected. Such sophistication is by no means limited to larger aircraft, either. One system, the Bendix BX- 200, stores all parameters for 10 waypoints, and can be preloaded with magnetic strips programmed away from the airplane with a pocket calculator. This system can even generate a "moving map" of the navigation situation which is displayed on the screen of the aircraft weather radar, yet the entire RNAV system, including a dual nav-comm system with autotune capability, can be panel-mounted in under a foot of stack space!

At the upper end of the spectrum are systems that not only provide RNAV, but VNAV, or *vertical navigation*, as well. Thus, they not only provide horizontal guidance to a waypoint, but they also tie into the encoding altimeter to provide vertical guidance as well. Waypoints can be crossed at preselected altitudes, and climbs and descents can be made at preselected vertical angles, something which will become increasingly valuable as fuel prices continue to climb. Even now, a VNAV is handy for general descent planning, not to mention those situations when the controller snaps, ". . . cross 13 west of Fillmore at or below FL 170, 10 east of Lake Hughes at 9500 . . . " The most advanced systems also have the capacity to store hundreds of waypoints in a (pilot-alterable) permanent memory which remains "alive" even when aircraft power is shut down, and these waypoints can be recalled either individually or by complete RNAV route. I won't go into the details of individual systems here; for our purposes, it is sufficient to say that they require a bit of practice to use, but that once you get used to programming them, you'll wonder how you ever got by without them!

Long Range Nav Aids

If your flying duties start taking you beyond the borders of the United States, particularly over water, you'll encounter some of the newer long-range nav aids. These, as much as the aircraft themselves, have made long-range and overwater flights feasible for corporate aircraft. As recently as 15 years ago, only the airlines and military had any sort of long-range electronic navigation capability. Other aircraft still had to depend on the same basic sextant and chronometer system that had served Captain Cook!

The first major breakthrough came with the introduction of *Doppler Radar* to smaller aircraft. This system is based on the principle of Doppler frequency shift, which states that a frequency generated by (or, in this case, reflected from) a moving object will be higher in proportion to the speed than that of a stationary object. The classic example is to stnad at a level crossing and listen to the whistle of a passing train; you'll hear it drop sharply in pitch as it passes your location.

An airborne Doppler set transmits four radar beams downward toward the surface; they're arranged in two pairs, front and rear and left and right. The frequency difference between forward and rearward beams gives you an accurate groundspeed, while that between the two side signals gives a sideward velocity (or, when combined with heading information, a drift angle). In a sense, a Doppler does the same thing you do when you fly pilotage—it keeps track of your position by monitoring your track over the ground.

Dopplers have some disadvantages, though. They're heavy, expensive, and require lots of maintenance, and they're not always 100% accurate. Operated over water, they "see" only the ocean surface, so surface wind, waves, and currents can throw them off. Over a completely glassy sea, they won't work at all.

The next big breakthrough came with the development of inertial navigation systems, whichuse delicate accelerometers on gyrostabilized platforms to measure aircraft movements. The heyday of inertial nav for airplanes in the medium twin class is probably already over, since newer systems provide similar performance at much lower cost and weight. Even today, with improvements in inertial systems, a decent one still will weigh more than 50 lbs and cost close to a quarter million dollars!

Nowadays, there are several major long-range navigation systems in use, and most of them are based on one form or another of low-frequency or very-low-frequency radio reception.

The low-frequency band is the location of the Loran-C system, which operates at 100 kHz, about half the frequency of the lowest NDB's.

"Chains" of Loran-C stations have been installed at various locations world wide, with more planned, and sooner or later Loran-C coverage will be virtually global. Compared to older Loran-A systems, which required a considerable degree of operator skill and training, the Loran-C units are extremely simple to operate. All you have to do is merely turn them on, initialize them with current starting position, time, and date to compensate for radio propagation differences), and you're on your way, with position displayed in terms of latitude and longitude. In airborne systems, waypoints can be entered (also in lat/long format), and the system will read distance and bearing to waypoints at any time, as well as driving the usual cockpit displays. This is true random navigation.

Further down the frequency spectrum are the VLF (Very Low Frequency) units. For these, coverage is virtually worldwide with relatively few stations, since signals at these low frequencies follow the curvature of the earth.

Two major systems are in use here, the U.S. Navy's VLF communications network and another U.S. Navy-derived navigation system—Omega. Each has its proponents , and each has its advantages and disadvantages. By now, most of the better VLF sets receive and use *both* signals, making the arguments somewhat academic.

Operation of virtually all VLF systems is very similar (and is similar to that of inertial units, too; in fact, sometimes a pilot has almost no way of knowing which system might be attached to the other end of the control head!) All must be *initialized* with GMT time and date to allow their onboard computers to insert appropriate corrections for diurnal variations in radio propagation, and the starting position must be loaded (some units automatically store their position when shut down, so they "wake up" knowing where they are. Additionally, a number of types incorporate a quartz clock module with its own tiny battery, so they're ready to go almost as soon as you turn them on if they haven't been moved while "asleep."

As with other types of unit, waypoints are usually keyed in as lat/long pairs, with outputs provided to drive cockpit displays and autopilots. Additionally, all sorts of additional information can be called up, such as actual winds aloft, distance and time between waypoints or from present position to any waypoint (either direct or along the planned route), and so forth.

One of themost sophisticated of the new VLF systems is Collins' LRN-85. In addition to all the features mentioned above, the system stores the locations of all the VOR's in the world and most of the airports of 4,000 feet or over by their three-letter identifiers, as well as by lat/long coordinates. Thus, if you were in San Francisco and wanted to fly direct to Kansas City VOR to start the ILS approach there, you could just punch in "SFO-MKC" and be on your way. Note that the system doesn't actually *receive* VOR's; it just knows where they are, to simplify flight planning and execution, and it'll take you right to a VOR station, but using VLF signals.

14

Flight Deck Management

Sooner or later in your career, you'll probably encounter a two-crew aircraft with captain and copilot. If it's sooner, you're likely to be sitting on the right; if it's later, on the left. In either case, let's close this little book with a few thoughts on cockpit managament and delegation of preflight and inflight duties.

Traditionally, one of the perks of captaincy has been the privilege of lounging int he pilot office while the faithful copilot preflights the aircraft in freezing rain, icy wind, blowing dust, or whatever. Having experienced this class distinction from both ends, I suppose it's inevitable. If you're a copilot, don't skimp your preflight; remember, apart from your responsibilities to the passengers, that you will be flying in that thing, too. By the same token, don't let the captain rush you through it if he's in a hurry—he'll be flying in that thing, too!

If you're the captain, of course, there could be better things to do than lounge around drinking coffee. This is a good time to zip down to the ol' FSS and really get the skinny on what the weather's going to be like, and once you're aloft, you'll earn the gratitude of your copilot, as well as teaching him (and possibly yourself) a lot, if you take the time to go over the weather in detail with him.

As far as inflight cockpit manner is concerned, I've flown with pilots ranging from studiously polite (" . . . the left engine is burning. Would you feather it, please, and get the fire extinguisher? Thanks so much . . .") to the traditional martinet (". . . Mr. Christian!! Bring that man aft!"). Somewhere in between is the happy medium in which things get done with a minimum of fuss and a minimum of friction.

Dividing the Duties

Probably the most important area is to decide from the outset who's going to be responsible for what. A classic story concerns the great Hollywood pilots, Paul Mantz and Frank Tallman, who were on their way back northward from a Baja California fishing trip with a Twin Beech load of screen celebrities. Enroute, they were forced to penetrate a significant line of thunderstorms, so they rigged the Beech for heavy weather and plunged in. For over an hour they fought their way through the storm; each time the airplane staggered in a particularly wild gust and fell off on a wing, or glowed luridly with lightning, they'd look hard at one another. the weather was truly hellish, but the airplane was stout, and it was commanded by two pilots of legendary ability, so the altitude and heading remained, if not constant, close to what was required. As Frank used to recount the story, "It was wild, and those movie stars were losing their lunch all over the airplane. Finally, we broke out and the airplane settled down. I looked across at Paul and said, 'That was as bad as anything I've

Table 14-1. Possible Division of Duties for "Pilot Flying" and "Pilot Not Flying" an Aircraft.

Situations	Pilot Flying	Pilot Not Flying
Before Start:	performs checklist items	reads checklist
Starting:	starts engines	monitors instruments
After Start:	checklist items	reads checklist
Taxi:	taxis acft, monitors radio	copies clearance
Pre-Takeoff:	checklist items checks radios	reads checklist sets up radios
Takeoff Roll:	steers airplane, applies initial power	sets exact power, guards throttles
After Liftoff:	calls gear, flaps calls climb power	operates gear, flaps sets climb power handles communication
Engine Failure:	flies aircraft determines failed engine calls feather as req.	attempts to restart or feather as req. checks acft. config.
Level-Off:	levels off calls cruise power	checks altitude sets cruise power
Cruise:	monitors nav situation monitors systems	handles communication monitors systems
Before Descent:	briefs PNF on descent and approach	checks dest. WX handles communications
Descent:	flies acft	handles radios; makes altitude callouts
Approach:	flies acft. by instruments ("eyeballs in")	monitors approach; calls out altitudes as required; looks for runway ("eyeballs out") and reports in sight.
Landing:	stays on instruments until visual landing assured, then lands	continues to monitor and call out airspeed, height, and sink rate. Sets radios for missed appch as required
Missed Approach:	flies acft	monitors missed appch procedure, handles communications
Shutdown:	checklist items	reads checklist

been through. I'm sure glad *you* were flying the thing.' Paul turned white as a sheet . . ."

Now, this story may or may not be true; knowing Frank, I imagine it was somewhere in between, but the moral is a good one. Nothing can cause more hassle, not to mention danger, than two crewmen not knowing what the other is doing.

The airlines seem to have developed the best concept, dividing chores between the "pilot flying," (PF), and "pilot not flying" (PNF). Either capatin or copilot can fulfill either roll. Table 14-1 show how I divide it up when I'm in a two-crew situation.

As you can see from the table, the role of each crew member is quite clearly set out. Note, however, that PF and PNF don't necessarily correspond to captain and copilot. In fact, they can be swapped back and forth during a flight, as long as each crew member knows what's expected of him in any given situation.

You'll also notice that the procedure is designed to free the PF for one major and all important task—driving the machine. What with the PNF setting up the radios, this could lead to confusion, except that good practice calls for the PNF to continually keep the PF appraised of the situation by explaining the setup each time a change is made: "The inbound localizer course is set on your HSI, the localizer is pegged, and there are no flags. The outer compass locator is on the ADF needle on your RMI, and the Rising Gorge VORTAC is on the omni needle; that's the missed approach fix. We're tracking outbound from Hog Jowl VOR on ly HSI; your heading bug is set to the missed approach heading . . . " and so on. Properly executed and with good cockpit discipline, and that does not mean that the captain has to carry a whip and chair, a technique like that above can make every flight, even those in rough weather, routine.

Of course, there are tasks that exist all the time that aren't called out in Table 14-1. Most of the time, both pilots tend to keep and eye on the general situation and on aircraft systems. I know it sounds dumb, but if you're flying as part of a crew and see something out of line, don't be quiet just because it's not your responsiblity. I know it sounds dumb, but it's happened!

Leg-Swapping

What about leg-swapping? That is, the practice of crewmembers swapping seats and roles on alternate flight legs. Personally, I'm all for it; some captains, notably those that feel that copilots should be kept in their place, oppose it. It's a real can of worms, and one guaranteed to cause controversy in the "letter to the editor" section of professional-pilot magazines. I feel that there's nothing wrong with an airplane having two captains aboard, as long as they both realize that only one is the real captain at any given time.

Index

A

Air-conditioning systems 61
Air-cooled engine 76
Airframe avionics 212
Air-source contamination 169
Autopilots 150, 208

B

Basic configurations 7
Beech King Air 190
- dead battery 204
- emergency procedures 201
- engine start 196
- fuel system 195
- interior 193
- landing 203
- preflight 192
- takeoff 199

C

Committed to land 23
Continental Tiara series 66
Controllability 8

D

De-Ice and Anti-Ice systems 54
- airframe 58
- props 56
- turbine engines 58
- windshields 56

Direct-Drive engines 65

E

Electrical systems 38
- generators 40

Engine failure 10
- at takeoff 10
- basic reactions 16
- critical 15
- dead foot 14
- dealing with 11
- feather 14

Engine operation 77
- climb 80
- cruise 81
- descent 82
- landing, taxi, shutdown 83
- starting 77
- takeoff 80

Exhaust-gas
- temperature indicators 75

F

Fire detection and
- extinguishing systems 62

Flap systems 52
Flight deck 220
Flight Director System 211
Flight log 129
Flight planning 99
- cruise altitude 101
- flight manual 104-127

Flying by yourself 131

Forced to go around 22
Fuel injection systems 74
Fuel loads 26
Fuel systems 28
fuel and vapor return 34
Fuel transfer pumps 30

G

Gas turbine engine 174
Geared engines 65
back-driving 65
lugging 66

H

Handling techniques 23
Heating systems 60
High level winds 134
Hydraulic systems 44
hydraulic pumps 45

I

Icing 138
IFR 128
Ignition systems 73
ILS 154
Induction air systems 73

L

Landing gear systems 47
Large aircraft 17
cruise situation 20
Light twins 17
Loading flexibility 25
Low visibility 152

N

Nav materials 129

O

Oil systems 72
Oxygen equipment 158

P

Performance 8
Physiology 156
Piper PA-31 Navajo 85
cockpit 89
cruising 96
engines 90
landing 96
preflight 86
pre-taxi 93
takeoff 95
Power steering and power brakes 53
Preflight paperwork 27
Pressure controls 163
Pressurization 157
emergencies 168
systems 160

R

Radar 145
Reciprocating engine 64
Reconfiguring for
maximum climb performance 13
Re-pressurizing at altitude 170
Rockwell Commander 840 42
RNAV 216
Doppler radar 218
waypoints 216

S

Stability Augmentation System 206

T

Thunderstorms and turbulence 145
Turbocharging systems 68
operating 70, 80
Turboprop engine 174
classes of 175
temperature 180
Turboprop power controls 183
Turboprop propeller systems 186
Turboprop starting 188

V

Vacuum air systems 35

W

Weight and balance 25

Z

Zero-zero approach 155